The Bee Walk;

being the romance and practice of beekeeping

by Amy Lisney

Northern Bee Books

THE AUTHOR   *Ross Studios*

The Bee Walk;

being the romance and practice of beekeeping

by Amy Lisney

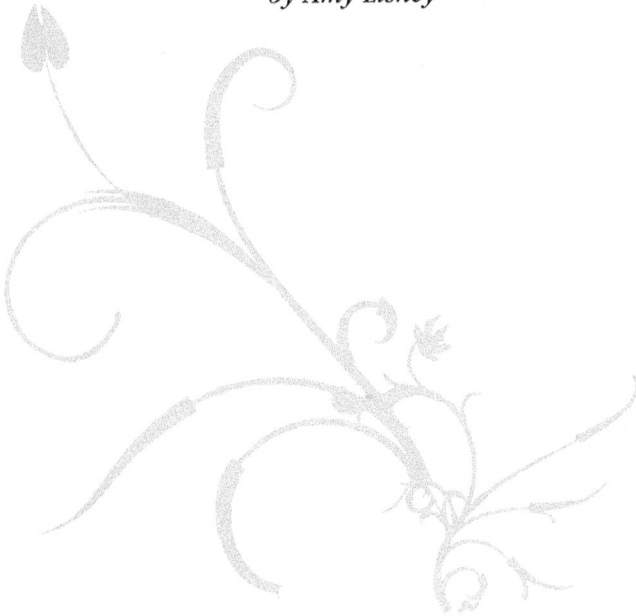

Northern Bee Books

The Bee Walk

© 2012 Amy Lisney

This is a republication of the title, first published by the author in 1953 in County Wicklow

ISBN 978-1-908904-22-5

Published by Northern Bee Books, 2012
Scout Bottom Farm
Mytholmroyd
Hebden Bridge
HX7 5JS (UK)

Design and artwork
D&P Design and Print
Worcestershire

Printed by Lightning Source, UK

To the Memory of my younger son,
LESLIE VERNON
Killed in Action in Burma, 1943,
who figures in
'The Swarm that was turned back,'
Chapter 5.

Jeremy Burbidge, is to be thanked for this reprint of *The Bee Walk; being the romance and practice of beekeeping* by Amy Lisney, published by the author at Bray, Ireland.

In 1953 Amy Lisney produced one of the literary gems of beekeeping, a book that stands alongside Tickner Edwards *The Bee Master of Warrilow* which was published a generation before in 1907.

Beekeeping has advanced immeasurably since *The Bee Walk* was written, with the scientific study of bee behaviour and bee diseases matched by educational programmes. Programmes organized by beekeepers Associations in the British Isles and Ireland as well as by central government.

I therefore commend *The Bee Walk* not as text book but as an enjoyable read.

Karl Showler,
Brecon.
Powys

CONTENTS

ILLUSTRATIONS

ILLUSTRATIONS

INTRODUCTION

'And God said, Behold, I have given you every herb yielding seed, which is upon the face of all the earth, and every tree, in which is the fruit of a tree yielding seed; to you it shall be for meat.'

GENESIS, CHAP. I, VERSE 29.

EVERY garden has many walks, but it is usually only one walk in a few gardens which is flanked with a row of bee hives. Such a walk is a constant source of joy and inspiration to the owner, sharing as he does in the life and mystery of his bees. And so back into the dim past of man's history, for the honey bee has been the most studied and the most written about of all insects.

More than two thousand years before the Christian era, the Fellaheen of Ancient Egypt took their hives on boats up and down the Nile in order to keep pace with the seasons in upper and lower Egypt and the flowering of plants on the river banks; the Romans too took their bees with them as they pushed down the Rhine.

At this time, however, man had little real knowledge of the bee but cherished many pretty dreams and fancies about it; observation and speculation continued and, according to Cicero and Pliny, the philospher Aristomachus spent all his time during fifty-eight years studying bees. Yet it was only as recently as 1688 that Swammerdam, a Dutchman, discovered the profound truth, namely, that the individual bee which rules and gives life to the colony is not a king as previously thought, but a queen. Then followed Réaumur, a Frenchman; Hüber, a Genevese Swiss; and Dzierzon, a German from Poland, who in turn added their contributions. These four scientists in fact set the foundation of our modern knowledge and practice; the glass observation hive being invented by

11

Réaumur, and other improvements by Dzierzon who also observed parthenogenesis in the bee.

Later improvements by an American, Langstroth, led the way to the modern hive, and the Bavarian, Mehring, provided the hive with sheets of stamped wax so that bees have only to follow the outline of the cell in order to complete its rudimentary walls, thus saving both wax and labour. A Venetian, Hruschka, devised the centrifugal separator so that honey could be extracted without damaging the fragile walls of the cell, and gradually, as methods of apiculture were improved, beekeeping became more and more popular; the veil and smoker gave a certain amount of confidence, with the result that the sting was no longer feared to the same extent.

In spite of modern developments airy fancies are still woven about the bee, but only direct observation, carried on patiently for many years, can reveal the truth. Some mysteries still remain partly unsolved, such as those, for instance, connected with the swarm; the nuptial flight of the queen; and the sympathetic vibrations linking the bee and the flower.

Darwin in his *Origin of Species* writes 'Thus I can understand how a flower and a bee might slowly become, either simultaneously or one after the other, modified and adapted to each other in the most perfect manner, by the continued preservation of all the individuals which presented slight deviations of structure mutually favourable to each other.' Here Darwin touched on the fringe of a great fact, namely, that embodied in this frail insect is exquisite mathematical precision; exact geometry is applied to the methodical and uniform construction of the fragile and diaphanous hexagonal cells, which are arranged in perfect order side by side, without interstices resulting in waste of space. In this way none of the minute and precious drops of nectar gathered so laboriously from the flowers, are wasted.

Nor have Darwin or others been able to explain the

mysteries of the antennae which receive messages of meteorological importance, and contain olfactory cavities which detect with such swift and delicate precision the call of the flowers.

Thus the bee carries on the work for which it is appointed, namely, the pollination of herb and tree blossoms and the provision of surplus honey for mankind; work that involves the organisation of a colony on the most perfect system imaginable, which staggers the imagination of man. So the mind of God is revealed in this fragile insect.

If I have winnowed a few grains of truth from the observations of a lifetime among my hives up the bee walk, then this modest contribution to the literature will have served its purpose; in any event it has been time happily spent giving me a casket of precious memories.

Finally, I should like to express my indebtedness to my son Arthur for assisting me in fusing into book form the records from my diaries over the years. Also to Dr. R. H. Barnes and Mr. A. Worth of Dorset for reading the manuscript; and to those who have very kindly provided the illustrations, for which they retain the copyright.

AMY F. E. LISNEY.

1

EARLY IMPRESSIONS

It is in the Spring, when the earth is beginning to rouse herself from the cold of winter, and the early snowdrops and crocuses peep out shyly from the shelter of their leaves that, coaxed by a burst of warm noontide sunshine, some of the most daring bees come out from their hives. These are the water carriers who go out to collect it all the day long, though at first only the brave few venture forth on this difficult, dangerous and exhausting work.

Later when the weather improves, the bees who seek the rich golden pollen from the anthers of the early crocus fly bravely home with either the precious drop of water, or it may be their little 'baskets' packed with golden nourishment; a dark cloud drifts across the sun; they become chilled; their wings are paralysed; they drop to the ground unable to rise or even crawl. Is there no-one to help?

When such casualties were observed I had learned from my brothers when a child gently to pick up and cup the little cold body in my hands and 'haw' on it. The experience of seeing the almost lifeless bee revive was so exciting. After a shuddering convulsion it would get on its legs and dust itself, first brushing the antennae with the forelegs and then raising up the abdomen and dusting it vigorously with long sweeps of the brushes on the hind legs. This completed the little bee would jerk its triangular head from side to side, and having decided that the last atom of dust had been removed would fly off like an arrow.

All very absorbing and fun to watch: how happy I felt. I was only stung once on the lip, but the pleasure in restoring life and activity made up for it.

Bumble or humble bees came in for first aid too, and revived with the warmth of my breath as they lay cupped in my hands. They make a most pathetic gesture when coming to, by raising up one of their forelegs as if pleading for mercy.

Some years later we often went to spend an afternoon and evening with an old lady who kept bees in skeps on stands along a sheltered walk in her flower garden. All summer there was great activity as the bees poured in and out, carrying stores for the use of the hives. I loved to watch the busy throng any afternoon I was allowed to go there, and so the summer passed; later there was talk of the honey harvest.

In early autumn when the skeps were well stocked with honey, they were weighed and the heaviest chosen to be smothered with sulphur fumes; I gathered all this from scraps of talk I heard. The bees who had toiled all through the sunny summer months to gather rich stores for the long winter were to be destroyed, so that man could plunder unmolested; in return for all the fragrant honey they had so carefully gathered they were to have a horrible death by smothering!! The thought hurt; I was too young to put in a protest that would be heeded.

This horrible slaughter would take place one evening when the bees were all murmuring happily in the skeps, the day's work over. Little pits were dug in the earth in which sulphur candles were placed, and the selected skeps placed over the pits and the candles lit. A dense acrid smoke rose from the pits and invaded the skeps, finding its way between the combs and thus stifling the unhappy bees whose frenzied wing beats were powerless against the deadly fumes: there was no escape.

A sudden stillness fell upon those thousands of wings which a short time ago were vibrating joyously; silent the monotonous chant that was audible some distance from the skeps in the soft early autumn evening. They were all dead.

The same evening the honey combs would be carried in on a

lordly dish golden rivers flowing thickly from them, and floating along slowly in the stream I would see a little lifeless body. On these occasions I found some difficulty in keeping back the tears, remembering how often I had warmed these small insects in my hands in order to have the pleasure of seeing them come to life again.

The old lady, beaming with satisfaction at the bountiful harvest, was congratulated on all sides by those who helped themselves from the dish with fruit and thick cream. When it came to my turn; 'what'!! 'you don't like honey'? I was too upset to eat; they would fly no more, those joyous harvesters building up their stores of food to last them through the perils of winter.

These experiences left lasting impressions on me.

Our First Hive

It was a red letter day when the first brand new hive arrived one morning, one of my brothers having previously announced that he was going to keep bees.

A few days later he came to me in a great hurry. 'I have bought a skep of bees from a neighbour to stock the new hive,' could I call out there one evening with the pony and trap and bring it back. 'I'm off to town on business for a few days—oh! bring a sheet and the carving knife with you'—all in a breath. While my head whirled with excitement I asked 'what's the knife for?' 'Oh! yes, I forgot' was the answer, 'just run the knife around the base of the skep to free it then lift it and see if its heavy, try the next and so on, Mrs. N. . . . said she would like me to have the best. When you do pick out the heaviest just spread the sheet on the grass, lift the skep on to it, tie the four corners up around and there you are; wait till it is dark, when all the bees will be in for the night.' It sounded simple enough. I was thrilled, to have bees of our own at last! how wonderful!! I rushed off to tell my sister who joined me in everything.

We were playing tennis that evening, a game we usually enjoyed till the balls were sodden with dew, but we eagerly postponed our sport in order to collect the colony of bees. Putting cloaks over our muslin frocks we set off greatly excited, but it was getting dark when we got to the little farm; Paddy came with us to look after the pony. We could just make out the 'skips' as Paddy called them, which were in a row all along a wooden plank and not far from the cottage door. I advanced bravely enough to pass the knife under the base of each as instructed, but found them firmly fixed to the plank as if they had been nailed on—later I learned that bees are very careful to make their combs secure and fasten the skep well to whatever it may rest on with a resinous gum called propolis, so as to keep out both wind and rain and prevent shaking. The propolis which secured the skeps was still soft owing to the heat of the day, but the knife kept sticking and was very difficult to manipulate. It was some time before I succeeded in freeing the first skep sufficiently to lift it from the stand, and I found it to be very heavy. The bees, objecting to the onslaught, became noisy and there was an angry hum. I felt hot needle-like pricks on my hands and face, and they buzzed in my hair. I went on to the next skep, it was not so heavy. I tried a third, and after it was released I decided that the first was much the heaviest and returned to it.

There was no mistaking the menacing roar now and the air was full of angry bees, but I knew nothing then of veil, smoker or carbolic cloth. Having decided on the first colony I spread the sheet, lifted the skep and quickly placed it in the centre then, gathering up the four corners and tying them carefully, I carried the skep back to the pony and trap; by this time it was quite dark.

I could not see anyone near and called as I made my way out of the garden. Lucklily there were arches and hedges, so brushing past them I left behind most of the bees that were buzzing angrily around. My sister joined me; she too had been

THE AUTHOR AT WORK IN HER APIARY (*Page* 11)   *A. A. Lisney*

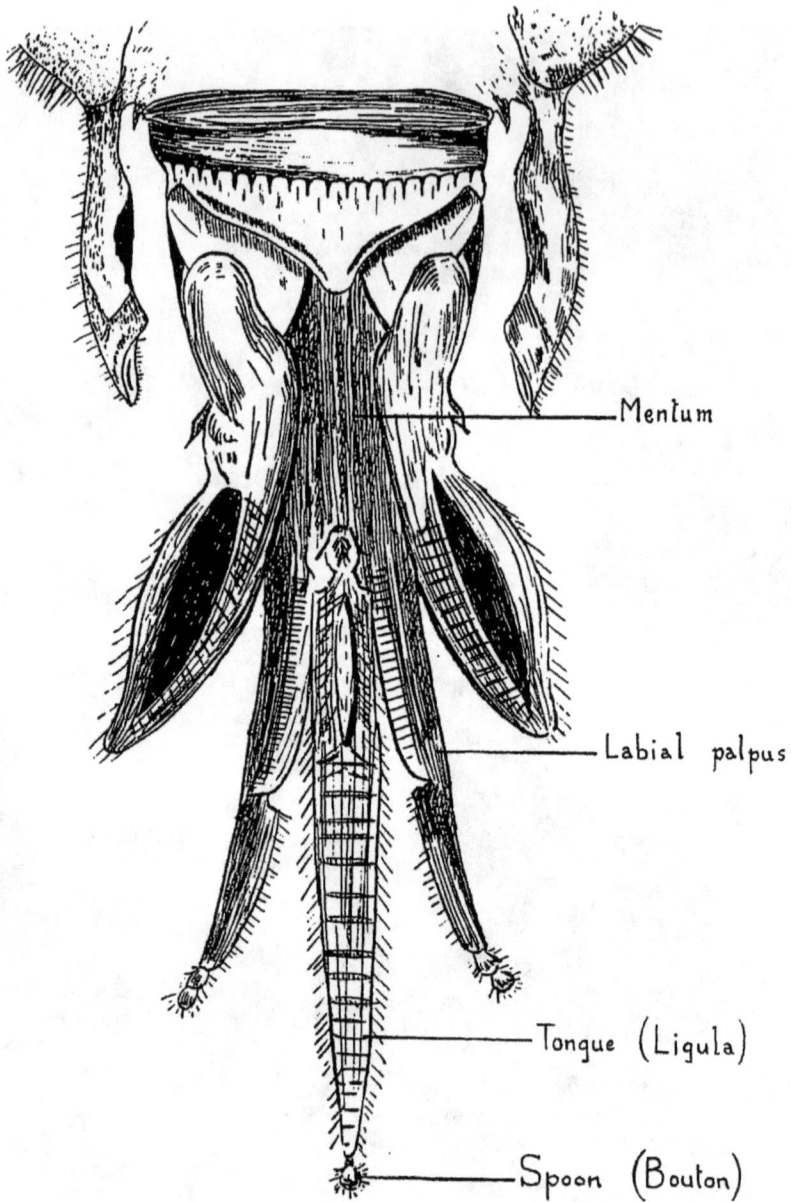

Mentum

Labial palpus

Tongue (Ligula)

Spoon (Bouton)

MOUTH PARTS OF HONEY BEE (*page* 26)

J. Waite

stung, but as she saw she could not help she had gone back and waited till I came along with the skep—and the knife; it is a wonder I remembered it.

When Paddy heard the bees—there were still quite a few in my clothes—he was very much disturbed. 'If any gets on the pony Miss we're done, he'll go mad, and throw us into the ditch, bad manners to them anyhow!!' In the twilight I could just see him beating the air with his cap. 'It's only a few' I soothed him, 'they are all tied up safely in the skep, let's hurry home.'

I do not clearly remember what my feelings were on the journey home; my sister and I supported the skep between us from which came an angry roar, but they had quietened down somewhat by the time we arrived. We carried the skep into the hall and I shall never forget my mother's face, glad to see us back but horrified at our appearance. I took the bees to an outhouse in the yard and when I returned my mother was picking the stings from my sister's face; she then proceeded on mine lecturing us all the time.

A friend had promised to come early next day and initiate us into the mysteries of beekeeping, but I had passed a very anxious night and imagined that all sorts of horrible things would be discovered when we opened up the sheet; would they all fly out and attack us? So little was my experience.

Gaining Experience

Early in the morning our friend arrived. Tall and commanding we felt he would have the situation well in hand and our spirits rose.

The new hive had been prepared beforehand and placed at the bottom of the garden, a wood in the background and a lake beyond. I had wondered why it could not be nearer the tennis court, but was told that the bees must have water near at hand; later I learned how very important water is to the life of the colony. Bar frames and comb foundation, all ready to be drawn out, had been fitted into the hive and over these had been placed a perforated zinc queen excluder.

Our friend had brought a small box and some turkey feathers with him. The box contained a square of muslin known as the 'subduing' or 'carbolic' cloth, which has a deep hem at one end wide enough to allow the insertion of a light cane about half-an-inch in diameter, and is damped with a weak carbolic solution before use.

While untying the sheet, our friend explained that bees become quite docile in the presence of carbolic fumes. Suiting action to the word he passed the carbolic cloth under the skep, and after a few moments reversed the entire skep so that it was upside down. He then placed an empty skep over it so that the open ends were together, and using two meat skewers with the ring ends joined he pushed the sharp ends, one into the edge of each skep, so as to make an improvised hinge. Then asking us to drum gently on the sides of the lower skep the bees, with a deep humming, began to move into the upper empty one. The march was slow at first, then faster until all the bees had transferred. The full skep was then laid on the ground mouth down after the skewers had been removed, while we examined the one the bees had just left. The combs hung like thick curtains varying in size so as snugly to fit the dome of the skep, and with just sufficient space between them to allow the bees to move about. In the centre combs there were eggs, larvae and unhatched brood in varying stages of growth. The cappings of the sealed brood were dark compared to the delicate ivory of those over the cells containing honey.

At this point our instructor told us to place the skep from which the bees had been driven, and which was full of brood and honey, on top and in the centre of the queen excluder in the new hive; then to pack it all around to prevent heat escaping before putting the lift in position and the roof on top. All was now ready for the final act. Giving the driven bees in

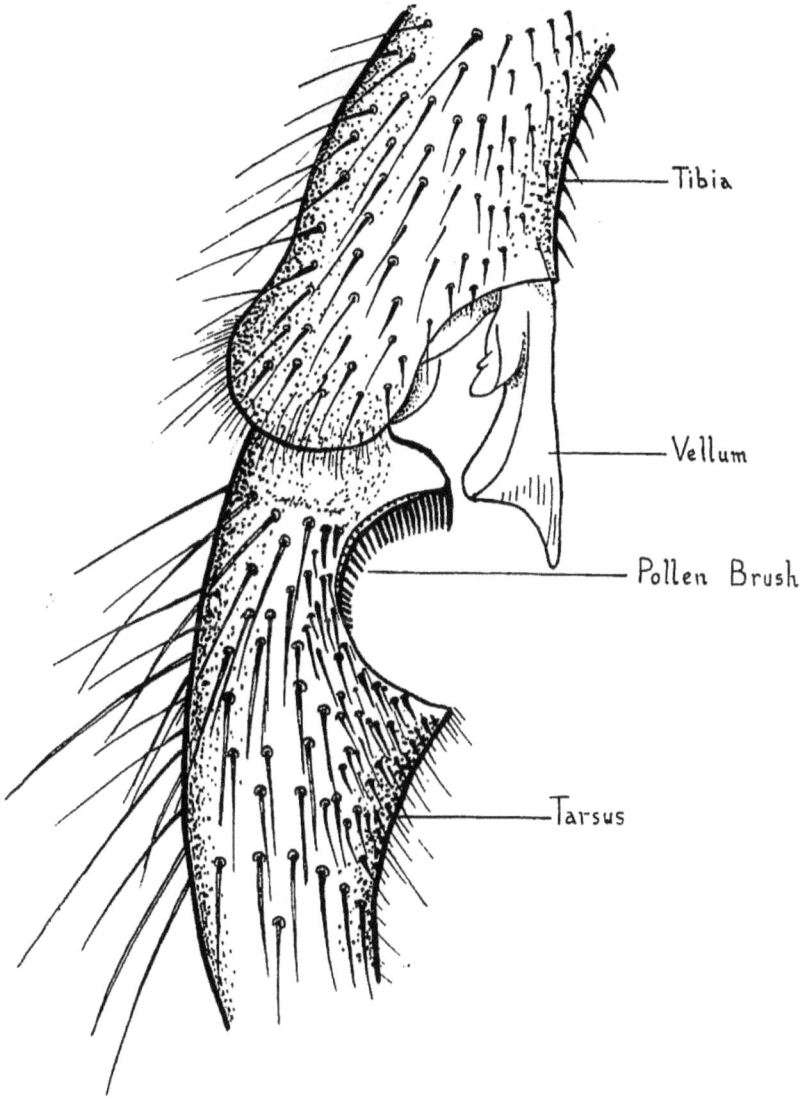

— Tibia

— Vellum

— Pollen Brush

— Tarsus

FRONT LEG OF WORKER BEE (*page 26*)　　　　　　　　　*J. Waite*

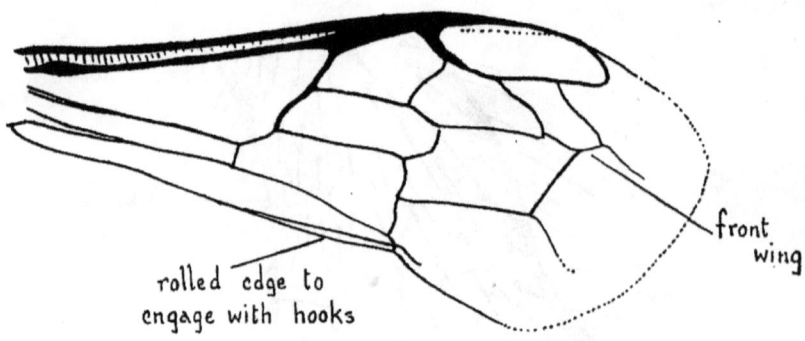

rolled edge to
engage with hooks

front wing

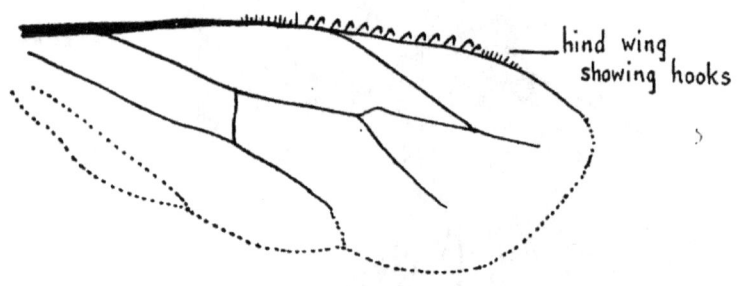

hind wing
showing hooks

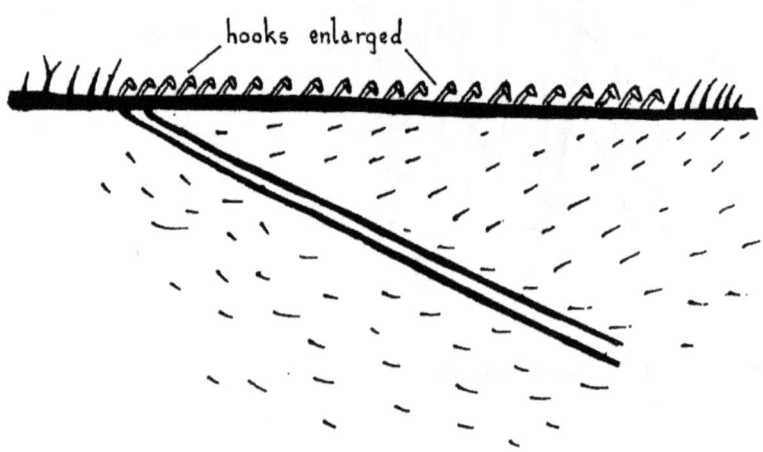

hooks enlarged

WINGS OF HONEY BEE *(page 27)*

J. Waite

the second skep a whiff of the carbolic cloth, they were deftly shaken on to a sheet which had been placed over a suitable piece of wood that sloped up to the hive entrance and rested on the alighting board. The bees began to move towards the entrance helped by a feather dipped in carbolic solution, but some flew up in the air. Suddenly an angry bee went for the instructor and stung him on the tip of his nose, and I regret to say that my sister and I went off into uncontrollable fits of laughter, no doubt due to the reaction after all we had been through since the evening before. Our friend smiled indulgently and made good the opportunity to tell us how to treat bee stings, 'a little of the carbolic solution'—here he dabbed his fast swelling nose with a cloth—and, as I made a sweep with my handkerchief, 'never' he said 'try to fight them but keep cool.' Fortunately we had no further trouble from the bees and were able to bend down in an endeavour to see the queen as she entered the hive. It is not easy to pick out a dark queen as she can easily enter the hive under cover of the workers, and so on this occasion it was left for me to see a queen later on.

The bees settled down to work and build up for the winter, with all the honey in the original skep to keep them going. The following year they were very strong and produced three crates of honey.

There was always some kind friend ready to assist with our difficulties, and one such neighbour offered to remove the honey for us. On looking back to that time not many beekeepers seemed to wear a veil when handling bees, which did not always end well. This particular neighbour went to our hive alone on a warm afternoon in July and, although he subdued the colony in some way, we saw him toiling bravely up the garden path carrying the crates with a cloud of bees after him. Regardless of ceremony he walked towards the tennis court while a game was in progress, and made straight for a military tent we always had on the lawn in summer time.

B

There was a table in this tent, but not knowing his objective everyone had fled to the tent for shelter. It was amusing to see the guests diving under the flaps in spotless flannels and muslin frocks as he strode in to put his load on the table.

Next morning our neighbour was unrecognisable, and had no little difficulty in explaining away his black eye and swollen face to those who did not know the heroic part he had played the day before.

FURTHER EXPERIENCE

I did not make any progress at beekeeping for some years. I left home and did not have a hive of my own till a brother sent me a colony of Cyprian bees.

At this time I knew very little of the various races of bees and was delighted with the gift, but I learned later that Cyprians are extremely nervous and irritable, smoke often having no subduing power over them. They will fly up in the air away from smoke and then come down to attack their owner, and the more smoke is used the more annoyed they become. These are in fact the most irritable bees ever brought from the East, so cross indeed that there is now probably not a single colony in the country, all having been entirely discarded.

I was removing my first two crates of honey from this hive one morning, protected by a veil but not using a smoker. The bottom of the lower crate had not had sufficient vaseline smeared over it to prevent the bees sealing it down to the hive, though as it was only the first week in July there should have been little possibility of this. I was obliged to release the crate with a hammer and chisel, and although the bees were furious I was not worrying as I had a veil. Suddenly I felt hot needle-like pricks on my face and neck, and to my horror found the veil had worked out from under my coat with the tussle to loosen the crates. The bees were over my face and in my hair.

Normally at this point I should have had the cane in one end of the subduing cloth and drawn it under the crates as I lifted them up, but with the weight of the full crates I could not carry out this manoeuvre, and anyway the hammering had annoyed the bees beyond control. I had to act quickly and looked around for the gardener who had been working nearby, but he had completely vanished. The potting shed was not far off and the door was open, so with all my strength I carried the crates into it, put them on the shelf and covered them up. Hurrying back to the hive I also covered it up, and returned to the shed where I sat down to think out my next move. I was covered with angry bees which had worked their way to wherever it was possible to sting me. I removed my hat and veil as they were serving no useful purpose at this stage, and wondered what it felt like to die from bee stings. After all I knew very little about bees or all this could have been avoided. 'Well' I thought 'there cannot be more bees outside than here in the shed,' so I opened the door and went swiftly down the garden, passing along under trees and arches so as to leave behind as many bees as possible before entering the house. As soon as Mary the maid saw me she exclaimed in horror 'Oh, Mam! all the bees are after you.' I went into the kitchen and put my head into a large basin of water; it was at least cooling. 'Mary,' I said, handing her a hairpin, 'do you mind taking a bee out of my ear?' She was very self possessed and had it out in a trice without it stinging me. She then proceeded to pick all the stings out of my face and neck, letting down my hair and clearing all the bees away. I began to feel better. 'Thank you, Mary, I'll go up and have a good wash and tidy.' I went up stairs in a sort of dream and wondered if I should ever come down again.

I sat down on the bed and felt wonderfully calm but ever so queer, something was going on all over me like waves surging through my body. I then began to tremble as if cold, and after a while waves of warmth passed over me. Presently I began to perspire from the top of my head to my toes and experienced a prickly sensation then, as I looked at my bare arms they became covered with a pale pink and white rash which spread over my entire body. I felt better, but continued to sit quietly for some time. I had little or no sensation of pain, neither was there any swelling except on my lower lip, yet I had received hundreds of stings of the most violent nature which even drew blood in many places. Having recovered sufficiently I went downstairs, got a basket and picked peas for lunch!

Previous to this I had read very few books on beekeeping and thought that bees were very much the same all the world over. My husband, considering it was time I had a good manual, bought me *The A.B.C. and X.Y.Z. of Bee Culture* by A. I. and E. R. Root and I became deeply interested in the subject. I procured a veil that extended to the waist and could not possibly work loose, and a Bingham smoker to replace the carbolic cloth which I never used again. I also purchased a queen of a more docile species, which taught me that it is possible to go through an entire season among the bees without more than an occasional sting. It is not that bees ever get to know one, they most certainly do not, but one gets to know *them*.

2

THE PHYSIOLOGY OF THE
HONEY BEE

HEAD—The head of the bee is flattened and triangular, being wider above and capped by the large composite eyes each consisting of from 3,000 to 6,000 facets set at different angles in the queen and worker, and from 11,000 to 13,000 in the drone. Three small convex eyes protected by a ring of hairs are set in a triangle between the composite eyes, the upper two looking right and left and the central one forward. This complicated eye arrangement may seem strange and difficult to understand, but it must be remembered that it is a fallacy to compare the senses and behaviour of the insect world with human experience, and assume that they are identical.

Research has proved that vision in many insects extends from the short wave light rays at the end of the spectrum invisible to the human eye through the range of human visibility, with the result that what may be darkness to us is not so to the insect.

Man has discovered radar to assist his vision in certain circumstances; an invention which makes use of radiation of longer wavelengths than those directly visible. Hence with radar man can 'see' in what normally is darkness to him, while the bee is provided with a vision which enables it to carry out exacting functions such as the construction of cells with exact geometrical precision in what, to us, are the dark confines of he hive.

The antennae are set upon the forehead like slender horns and consist of a number of joints working one upon another, extensible, flexible and always on the move. What wonderful mysteries lie hidden in these graceful appendages which science has not yet been able to fathom; that they receive messages in a language we do not understand there can be no doubt.

The tongue of the bee is half the length of the body when fully unfolded, and turns back upon itself so that the tip lies in the mouth when not in use. As soon as the tongue enters the chalice of the flower it slides forward and stretches to the utmost so as to reach the nectar, and at the same time the sides fold towards the centre along its whole length so that the tongue becomes a kind of tube through which the bee is enabled to draw up liquid.

THORAX—This is the front portion of the body corresponding to the chest, to which the legs and wings are attached. In addition to walking and cleaning its body, the legs of the bee are used to carry pollen, handle the wax scales and serve as the seat of some of the sense organs. All six legs are made up of several jointed sections which give independent and considerable freedom of movement.

The five principal divisions of the leg are given anatomical names; the coxa, that is the part where the leg joins the body, followed by the trochanter, femur, tibia, and tarsus from above downwards. The legs are covered with hair and spines, and of special interest are the dense bristles of the pollen brush with which the bee removes pollen or other foreign particles from the head and body. A special mechanism on each front leg is called the antenna cleaner. This is a crescent shaped indentation on the inner edge of the upper part of the first or elongated segment of the tarsus, the edge of which bears a single row of tiny stiff spines arranged like the teeth of a semicircular comb similar to that which our grandmothers used to wear. At the lower edge of the tibia a somewhat flattened projection called the vellum is located which hangs above the notch when the leg is bent at the joint between the tibia and the tarsus, thus covering the indentation. When the bee wants to clean the antennae it fits the flagellum, that is the whip-like part of each antenna, into the notch and bends the leg at the joint, thus catching the antenna in the indentation where it is held by the projection. When the antenna is pulled the teeth of the notch remove foreign particles adhering to one side of the flagellum, while the projection cleans the other.

The wings on each side of the bee are attached separately to the thorax, the front and hind wings moving in complete unison as, when in flight, they are locked together. The mechanism which makes this possible consists of a row of tiny hooks located on the fore margin of each hind wing, and a rigid narrow fold on the hind margin of each fore wing, so that when the hind margin of the fore wing is drawn over the fore margin of the hind wing the hooklets catch the fold and lock the two wings together.

Though fine as gauze the wings are stiffened by numerous nervures and are used to do more than bring in the harvest. The bees often stand at the entrance to the hive and vibrate their wings so quickly that they can hardly be seen, thus extracting the impure air from the hive and allowing fresh air to flow in. Thanks to this ventilating action, in addition to scavenging carried out by the bees, the hive is kept clean and wholesome. I should like to stress that these duties can be and are carried out by all the worker bees in a hive from time to time, and are not the sole prerogative of individual bees as some maintain.

ABDOMEN—Though the abdomen of the bee has no appendages as have the head and thorax, nevertheless it bears two most important organs, those containing the wax glands and sting.

There are eight wax glands in all, four on either side, which consist of specially developed cells of the skin on the ventral surfaces of the middle four of the six abdominal segments of the worker. The wax secreted by these glands is discharged through minute pores in the ventral plate of each segment concerned, and accumulates in the form of little scales in the pocket above the underlapping ventral plate of the seg-

ment next in front. How the bees produce wax is described
below.

Every worker bee has a body composed of six visible rings
or segments, one sliding into the other telescope fashion. When
the bee is full of nectar these segments slide out and the
abdomen is elongated considerably beyond the tips of the
wings. At the approach of winter the abdomen of an unladen
bee is very much drawn up and the insect appears much
smaller.

All bees have four bands of soft bright coloured down, one
on each of the four middle rings of the body, but none on the
first or last. These bands are very bright on young bees, and
give the Italian species a most attractive appearance ; the
labours of harvesting soon wear away these pretty fringes.
The abdomen also contains that complicated organ the sting
which is discussed in detail later.

A phenomenon which I have observed over many years is
that when hiving a swarm some of the bees line up facing the
entrance heads down abdomens raised, and wings locked and
vibrating at considerable speed. On the last segment of the
abdomen appears a tiny white crescent, now known as the
Nassenoff scent gland, from which, as I have always been
convinced, a delicate musk-like fragrance is exuded by
means of which they inform all the other bees in the swarm
that the queen has entered the hive. I am also of the opinion
that the same mechanism is used by scout bees when guiding
a swarm to the newly chosen abode, and on other occasions
when the bees wish to indicate the presence or safety of their
queen.

Aristotle declared that wax was made by the flowers; two
thousand years later man still believed that bees obtained
their wax ready made in this manner, and that in some
mysterious way the flowers were able to manufacture it from
pollen. In the eighteenth century Réaumur endeavoured to
discover how pollen, the crude wax as it was then called, could

NASSENOFF SCENT GLAND EXPOSED (*page 28*)   *H. Teal*

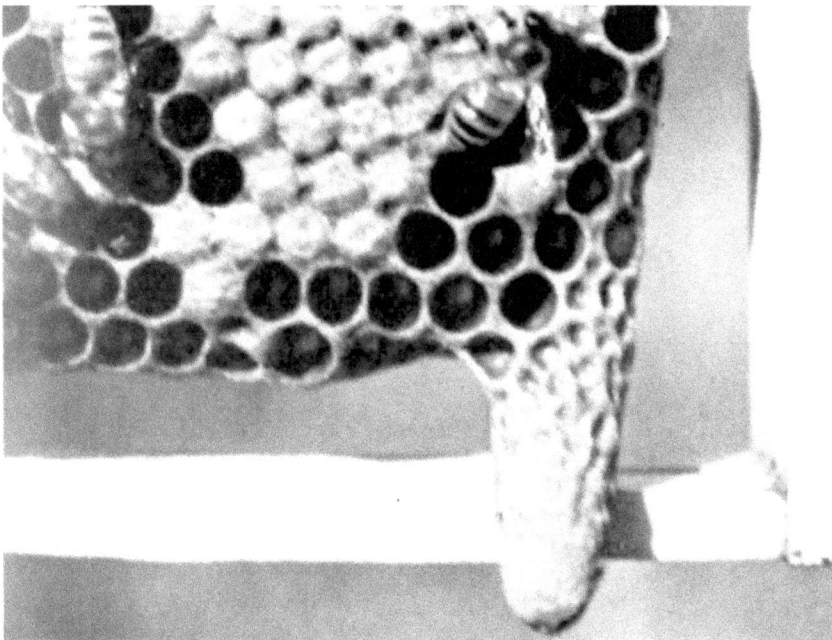

SEALED QUEEN CELL ON BROOD COMB (*page 68*)   *R. V. Roberts*

UNDERSIDE OF BEE SHOWING WAX SCALES (*page 29*)   *O. Lloyd*

WORKER BEE (*page 30*)   *C. J. Butler*

WORKER BEE CARRYING POLLEN (*page 31*)   *H. Teal*

QUEEN BEING FED BY NURSE BEES (*page 36*)　　　　　*R. V. Roberts*

possibly be transformed into actual beeswax; he made endless tests, but was always disappointed. Of one thing Réaumur was positive, namely, that the body of the bee was the only crucible in which wax could be prepared, but in exactly what way remained a mystery to him.

It was the blind Swiss, Francois Hüber, assisted by his faithful servant Francois Burrens, who discovered early last century that wax is a secretion produced by the bees which emerges as delicate scales from under the abdominal plates.

Honey and beeswax were used in Egypt thousands of years ago for embalming, and for making the coffins airtight. The odour of the wax would attract bees, and no doubt this gave rise to the belief held by the Ancient Egyptians that they had some important and mysterious connection with the dead.

From early times candles made of beeswax have been used in all churches, some of those provided at Easter weighing as much as sixty pounds. In former times the Church did not use candles made of paraffin or any of the mineral waxes, all of which give off an offensive greasy odour while burning whereas candles made of beeswax leave a delightful perfume. Moreover, the burning of mineral wax causes a deposit which injures pictures, while beeswax mellows and preserves them. The resistance of beeswax to heat is astonishing; it will endure a temperature up to 140° Fahrenheit before it melts.

DEVELOPMENT OF THE HONEY BEE

The honey bee, in common with other members of the insect world, develops through the stages of egg, larva, pupa and perfect insect; the first three of these stages in the comb being called brood.

The egg is cylindrical in shape and is laid in an upright position, being held in place by a gummy secretion. It is one-sixteenth of an inch long and slightly curved, thicker at one end than the other with a network mesh on the surface. It is always fixed so that the developing larva has its head downwards and as in any other egg it contains a nucleus with a surrounding layer of protoplasm, and a store of yolk as food for the embryo. On the second day the egg is leaning at a sharp angle while on the third day it is lying at the bottom of the cell. At the end of the third day the embryo, which has been developing and feeding on the yolk of the egg since it was laid, breaks through the shell and the newly hatched larva, just a piece of nakedness and little more than a stomach, eagerly awaits to be fed by the worker bees. The larva is fed for from five to eight days longer, when the mouth of the cell is sealed over. When fully fed the larva spins a cocoon and becomes transformed into a nymph or pupa, which later emerges as an adult bee.

If a queen is to be reared sustenance throughout the larval stage will be the rich brood food, that is the secretion from the lateral pharyngeal glands in the head of the nurse bee, which is only fed to the young worker larvae for the first few days; drone larvae on the other hand, are fed on nectar, honey and pollen.

The number of days passed in these early stages for queens, workers and drones may be summarised as follows:—

	Egg	Grub or Larva	Pupa in sealed cell	Emergence
Queen	3	5	7	15
Worker	3	6	12	21
Drone	3	8	14	25

WORKER BEES

All the work of the hive is done by the worker bees, which are imperfect females and are produced from eggs laid by the queen; the same eggs can be made to produce queens when the larvae are specially fed. Worker bees produce the wax which is secreted by their bodies; build the cells; fill them with honey; feed and nurse both the queen and worker larvae; collect pollen and propolis, a resinous substance which they find on trees and which is used for closing chinks and irregularities in the hive; defend the hive from enemies and keep it clean and well ventilated.

In time of necessity each worker bee can take on any of the work mentioned above, and during a honey flow all can become harvesters and respond in a body to the call of the flowers, leaving only the required number in the hive to look after the brood. It is this ability on the part of honey bees to take over any job in the hive at any time that gives so much purpose to their movements, which are instinctively organised with one object, namely, the harmonious prosperity of the colony.

When collecting nectar the bees do not begin work very early in the day. The caress of the morning sun must dry the heavy dew from the flowers before the bees can collect the drops of nectar concealed in the chalices; but those who act as water carriers are abroad early. Water is carried all day long to the hives where it is used for mixing the food given to the larvae. The water carriers are particularly attracted by the tiny crystal pools of dewdrops such as occur in the hollows of cabbage and lettuce leaves; safe hunting in summer time, but the frosty dew of early winter and spring is a real peril as many bees get chilled and fall among the leaves to perish. It is dangerous work if they have to seek these precious drops of water in running streams as a sudden gust of wind may hurl them to death by drowning. These water carriers do, in fact, perish by the hundred.

It takes fifteen thousand trips to and fro to collect the sixteen to eighteen ounces of water which a well populated hive will consume in a single day during the months of April and May, when the brood nest is expanding most rapidly.

Shallow dishes of water left out in a sunny sheltered position and often renewed, will attract a mass of eager drinkers. These should be filled with shells or small stones so as to give safe footing to the bees and thus save many a precious life. It is only in times of excessive drought, or neglect by the beekeeper, that bees are forced to visit dirty pools.

The production of wax is an exacting process which occurs at the heart of the hive where the temperature is high. The bees form themselves into a chain which extends between the two ends of a brood frame, the centre of the chain being just off the floor. Chain after chain is thus formed, the bees clinging to each other by their fore and hind legs; crossing and recrossing in loops and meshes they hang like a living curtain from end to end. Between the frames the temperature of this animated fabric rises to over 90° Fahrenheit. The bees have feasted on honey and appear to be performing some extraordinary and mysterious rite, and so much in a trance do they seem that one can lift up this curtain gently and let it fall into place again apparently without disturbing them. For twenty hours at a time they are held in this ecstatic trance, and the high temperature created favours the secretion of wax which appears from the wax-producing segments on the under side of the abdomen. These pearly discs are of rare beauty rather like the tiny delicate scales of a fish, and when they appear the bees dexterously pass them one by one to their mandibles so that they can be masticated with saliva and made ready for building the comb.

The life of the worker bee during the height of the season is approximately six to eight weeks, but in exceptional circumstances it will live up to eight months. The length of life depends on the amount of work done.

Drones

Drones are the male bees, hatched early in summer to provide suitors for the young princesses. They lead an idle life. The drone does not possess a sting, his legs are short and his tongue even shorter than that of the queen. His wings are very powerful, and although his brain is less developed than that of the worker his senses are wonderfully acute; the antennae, which consist of thirteen joints as compared with twelve in the worker, contain ten times as many olfactory cavities as those of the queen, and seven times as many as those of the worker. The drone is endowed with a delicate sense of smell and, according to Cheshire, his composite eyes possess more than thirteen thousand refracting facets which give amazing powers of vision. The drones leave the hive during the warmest part of the day, and require fresh nectar for their daily food which they may drink freely from the cells of the honey comb, although the worker bees also feed them at times.

As soon as the honey flow is over the workers turn the drones out of the hive to perish, unless there is no queen present.

THE QUEEN

The queen is much larger than either a worker or drone, and is normally the mother of every individual in her colony. She has a long downy body when young and her wings are short in proportion to her length. She is provided with a sting which is used both as an ovipositor and a weapon against rivals for the supremacy of the hive, but is not barbed like that of the worker. The queen can easily be distinguished if looked for early in the season, and it is a good plan to make sure that she is present every time the hive is opened on a warm sunny day from the end of April onwards. Care must be taken at all times, however, particularly when spending the time necessary for close examination of the combs, not to expose the brood to cool winds as it will become chilled and die, so retarding the development of the colony.

The virgin queen usually leaves the hive on her location flights in from three to five days after emergence from the cocoon, before taking her nuptial flight. On her return to the hive after mating, she will normally remain in it for the remainder of her existence except when swarming occurs. The one mating may suffice for her entire life, and three years is the longest period I have known a queen to live.

Although a queen must be mated in order that she may lay

the eggs which produce workers, eggs which produce drones are not fertilised being the parthenogenetic product of the queen herself. She has the faculty of laying at will eggs that produce either drones or workers, and at the height of the laying season produces from 2,000 to 3,000 eggs per day. The queen begins to lay in the warm centre of the brood nest each spring when the sun has begun to climb its way to the zenith; when early buds are bursting and a few timid flowers are opening their corollae. At this time the bees feel the stir of renewed life and are attracted by the scent of the early flowers. The queen is fed by the nurse bees with a special food from the lateral pharyngeal glands in the head so that she may be induced to lay, in anticipation of the thousands of workers required to bring nectar and pollen to the hive.

The development of the queen is one of the many wonderful episodes of insect life. An egg producing a queen in no respect differs from one which in the ordinary course of events would result in a worker. If, however, the colony wish to rear a queen a larger cell is constructed and a fertile egg placed therein. Queen cells are prepared in response to the swarming impulse; to supersede a failing queen; or, of course, following the loss of a queen.

It is important to memorise the different and progressive stages in the development of the hive bee, which have been referred to in detail earlier, so that if the queen is found to be missing from the hive, or if it is suspected that a swarm has got away, it is possible to date the time of their departure by the brood stage reached in the comb.

BROOD FOOD (Royal jelly)—As stated above, an egg pre-destined to produce a queen is usually laid in a large specially prepared cell, and throughout the entire larval stage brood food, that is the secretion from the lateral pharyngeal glands of the young nurse bees, is lavishly supplied during the four days of the feeding stage; the larva simply wallows in it.

So swift is the growth of the larva that, according to data

quoted by Dr. E. F. Phillips, the relative weight increases from 0·100 to 157·642 milligrams within the short period of five days.

Brood food as produced from the glands of the bee is of a creamy colour and texture. The residue of this food which usually remains in the cell after a young queen has emerged, is of a jelly-like consistency and the colour of amber with a streak of dark orange through it; it is firm to the touch, has a rich flavour, but no detectable smell. Hence the common term 'royal jelly'.

An analysis of brood food is:—

Proteins	...	45·15
Fat	...	13·55
Sugar	...	20·39
Water	...	20·91

There is just that difference in the food which on the one hand is responsible for stunting the growth of the sexual faculties in the worker bee, and on the other producing fully developed queens.

Species of bee used for honey production

The honey bee as we know it belongs to a large group of insects known as the *hymenoptera*, and although man has been interested in honey through many centuries he has found only a few races suitable for its production.

In the British Isles the Italian, common black bee and the hybrid, which is a cross between the two, are those most commonly used for honey production. Some years ago when the plague of Isle of Wight disease was at its height, Dutch bees were imported in the hope that they would prove resistant and replace the colonies decimated by the scourge. This, however, was not successful, the Dutch bees proving to have even less resistance to the disease than the others.

THE ITALIAN BEE—In my neighbourhood the Italian bee is superior to any other kind, and is used extensively by large table honey producers throughout the world; it is a more prolific breeder than the black strains.

Hives of pure Italian bees can be opened at almost any time without disturbing the cluster; they are very docile and if all their wants are anticipated seldom swarm. They also take very readily to the supers and are splendid foragers. I have heard recently that Brother Adam of Buckfast Abbey, Devon, has by selective breeding, even a better strain and limited numbers of queens are available from him.

For profitable beekeeping the pure Italian or Buckfast strain is, in my opinion, a bee that can be relied upon to give really satisfactory results under most conditions. They can be handled without smoke or veil though personally I always wear the latter, but have never worn gloves preferring elbow length sleeves.

When the hive is opened and the combs lifted out for inspection, the Italian bees remain quietly in place and do not rush wildly about and pour over the sides as the black bees do. The Italian queen will even proceed with her egg laying activities while the frame is held in the hands for examination, and her movements are worth watching. She dips her head into a cell and if it meets with her approval she turns, lowers her abdomen into it holding on to the sides, and deposits her egg. Often after carefully investigating a cell it is not approved and she moves on to another.

I have not noticed that bees pay their queen as much homage as is generally supposed. She will push past groups of bees to reach a cell and although it is obvious that they are quite conscious that she is near, they take little or no notice of her until she is fed, when they draw around her in a circle. Should, however, the queen be picked up and removed, the bees very soon know that she is missing and the hive is filled with a definite note of distress. They run hither and thither, out on the alighting board, up the sides of the hive; coherent work is suspended and all is chaos. When she is replaced

TYPICAL QUEENLESS COLONY (*page 36*)   H. *Teal*

DOUBLE GLAZED LID.

CORRUGATED TIN OR ZINC.

WIRE GAUZE

DOUBLE GLAZED LID.

CORRUGATED TIN OR ZINC.

WIRE GAUZE

METAL TRAY.

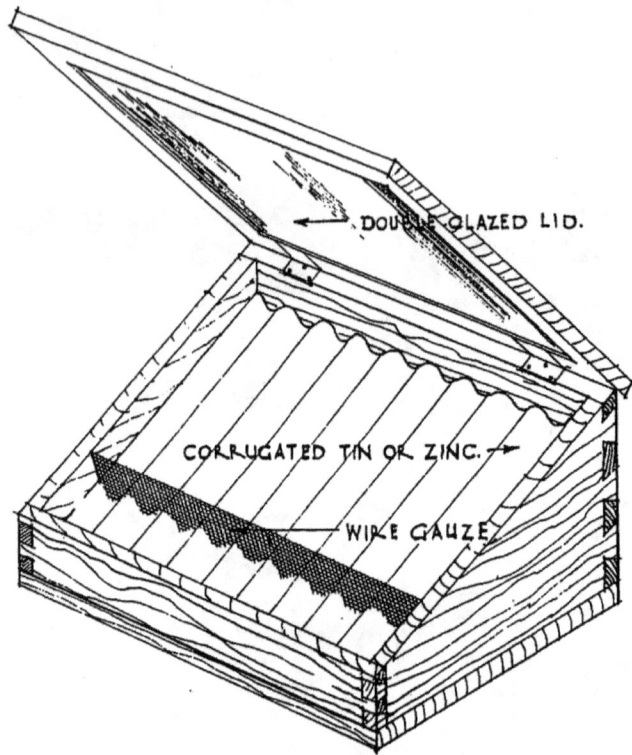

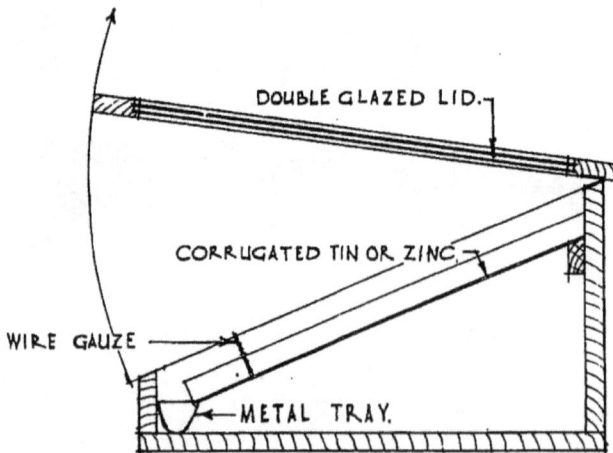

SOLAR WAX EXTRACTOR (*page 52*)   *G. Welch*

the joyful news of her return immediately spreads through the hive which vibrates to the musical chant of pleasure.

The Italian bee appears to stand congestion better than the black species and is thus not given to swarming if all the requirements of the hive are anticipated.

THE BLACK BEE—Black bees of native strain are irritable and very nervous, so that when the hive is opened they run like a flock of sheep from one corner to another in confusion, boil over the sides and hang thickly from the frames as they are held up for inspection finally falling in masses on the ground and crawling up one's clothes. They have a most annoying habit of following the apiarist from hive to hive and poising on the wing before his eyes just outside the veil. The strain of black bee I used in the past were inveterate robbers, so that during the feeding season great care must be taken to protect weak hives. The queens are difficult to find as they run and hide as soon as the hive is opened, and sting if caught.

A point in their favour is that when the hive is moved from the original site they will find their position much more readily than the Italians, the returning bees going from hive to hive until they discover their own.

CARNOLIANS—These bees are very gentle, but not more so than pure Italians. They are larger than the black bee which they slightly resemble and are excessive swarmers, which alone makes them very unsuitable for honey production. There is one very good point in their favour in that they deposit very little propolis.

CAUCASIANS—The Caucasians are also very docile, but in my experience their honey gathering qualities are inferior to other species. The most serious objection to them, however, is their propensity to swarm, and they are inclined to use rather a lot of propolis.

BANAT BEES—This species comes from Hungary. They are gentle and not so much given to swarming as the Carnolians, and build up very rapidly in spring. The queens are a very

C

handsome dark tan colour and when crossed with pure
Italians their offspring are very gentle, even after several
generations, unlike a first cross between Italian and black
strains which are often difficult to handle.

TUNISIANS—Natives of North Africa, this is a black race which
has been tested to some extent, but it has not been possible
to establish any claims in their favour and they are irritable.

CYPRIANS AND HOLY LAND BEES—These are the most nervous
and most difficult of all bees and smoke has little power over
them, in fact the more they are smoked the more enraged they
become. No garden or lawn is safe where they are kept as they
seem to lie in wait for the unwary who venture near their
hives. For this reason they have been entirely discarded.

EGYPTIANS—These are said to be the most beautiful species.
It is named *Apis fasiata* by entomologists and has been culti-
vated for thousands of years by the Egyptians, being probably
the first species used by mankind for domestic purposes.

In the time of the ancient historian Herodotus, apiaries were
seasons in upper and lower Egypt. Inscriptions on tombs
show the practice in use at least 4,000 years ago and indicate
that the honey bee was highly revered by the people of that
age.

GIANT BEES OF INDIA—The species *Apis dorsata* have their
home in the Far East and build huge combs of pure wax often
five to six feet long and three to four feet wide, which they
attach to overhanging ledges of rock or to large branches of
lofty trees in the primeval forest jungle. They do not construct
larger cells for their drones than does our hive bee.

There was an attempt in 1880 to introduce this species to
California, but the experiment failed and it was realised that
any climate, apart from that pertaining in the jungle, would
be fatal to them.

By way of diversion, it is interesting to note the wide
variation in the life histories of the bee family generally, and

the man who knows how to use his eyes in the garden may observe much of interest.

The solitary bees are obviously unsuitable to us for honey production. The smallest of these docile little insects build their nests in holes which they scoop out in sunny sheltered banks or old walls. The queen can lay eggs producing at will offspring of either sex. The young bees do not hatch until the following year, when spring is well advanced.

The leaf cutting bee lived permanently near the verandah of our home, and was the source of much interest and absorbing attention every summer. All requirements were near at hand, and the leafy thimble shaped cells were built where the pointing between the bricks on the walls had become loose enough for the cavity to be enlarged. There was every variety of shrub to choose from, but it was from the leaves of the wisteria which grew along the verandah that this bee cut with mathematical precision the ellipses to form the sides and bottom of the cell, and the round discs for the lids. In order to cut a piece, the bee stands on the upper surface of the leaf. The detached portion is held with the front feet during flight and placed in the position it is to occupy. The lower end of the portion of leaf is bent into concave shape so as to form the bottom of the cell, and any spaces are covered over by additional pieces. When the bee is satisfied that all is ready she goes in search of nectar and pollen which she kneads into a cake; this is placed in the bottom of the cell and on the centre of it the egg is laid. The cell is then neatly capped with a disc which is cut from the more strongly veined portion of the leaf.

During two memorable summers we were visited by the mason bee; it was of a light grey colour and built its cells on the corner stone of the yard wall, using as mortar very dry clay and a little sand mixed with saliva. Its whole body vibrated as it scratched with the tips of its mandibles and raked with its forefeet to detach particles of earth and grains of sand. It worked with such zeal and was so indifferent to passers-by that we had to exercise the greatest care lest we trod on it. The result of this labour was a cluster of half a dozen cells, very rough on the outside. During the second summer we observed one which built its little dome of cells on a stone in the rockery.

Bumble bees, although larger than the hive bee, live in much smaller colonies, making pretty little domed nests of moss on the ground in the fields, or in holes in grassy banks. They have a sting which is not barbed though very painful, but they are not easily annoyed. Towards the end of the season their larvae are fed on special food to produce queens to carry on the race. The queens, when fertilised, hibernate through the winter, burying themselves in mossy banks or even in the heart of masses of aubretia or saxifrage in our gardens, from which I have seen them emerge on a glorious day in March.

Wasps are very easily aroused to aggression and can sting time and time again as their weapon is not barbed. It is not necessary for a wasp to take hold of its victim in order to sting as it can do so while flying. The sting itself is not poisonous, but does carry infection when the insect comes in contact, while feeding, with contaminated material such as kitchen refuse and rotting fruit.

In April the queen wasp begins her hanging nest, or it may be a nest in a hole in the ground, laying as she builds. The first batch of eggs hatch out very much quicker than those laid subsequently, and these early workers take over the task of building and feeding the queen, who confines herself to laying.

When the days are beginning to draw in and summer is on the wane the wasp larvae are fed in order to produce queens to perpetuate the species. We see the workers with their powerful mandibles cutting off tiny joints from the meat in the butcher's stall, dexterously catching flies and snipping off their wings and legs and also collecting caterpillars from our gardens. All this varied meat diet provides the equivalent of brood food which the hive bee supplies to its larvae in order to produce queens. It is only for this purpose that wasps use meat, the normal food for their larvae being nectar.

Only fertilised queen wasps hibernate through the winter. In a nest built in the ground, which was dug out whole, we counted hundreds of queens in their nymphal state. In the late autumn most of the nests have been destroyed by rain and storm and the workers, having nothing to do, spend the rest of their lives robbing weak hives, pilfering jam and other sweets and feeding on decayed fruit. They have nothing to live for and become pests.

3

EQUIPMENT FOR THE
BEEKEEPER

THE HIVE—There are many types of hive listed for sale by reputable dealers. To name a few of the most popular, there is the C.D.B., the W.B.C., the Commercial and the National, any of which will with care last a life-time if constructed of the very best material and lock jointed throughout. The C.D.B., National and W.B.C. hives hold brood frames 14 × 8½ inches, while the Commercial holds larger frames, 16 × 10 inches. In England, however, modern beekeepers have tended to use single walled hives of simple construction, made from unpainted western red cedarwood.

In the C.D.B. for instance the floor board is a moveable wooden stand with four stout legs, the sloping portion in front forming the alighting board which projects well beyond the hive entrance. It is not necessary to have a ventilator in the floor as this upsets the method of ventilation produced by the bees themselves.

The brood chamber or body box comes next and rests on the floor board with the side and back walls overlapping it, and in order to provide an entrance for the bees a space is left between the lower portion of the front wall and the floor board. This part of the hive is the permanent home of the bees, and in it are suspended the bar frames in which the brood is reared and the winter food supply stored.

In Ireland the 'Abbott' bar frames are the type usually used. These are machine made with the greatest accuracy, and kept at the correct distance apart by the shoulders at each end of the upper bar. Today in England there is a tendency to use the self-spacing Hoffman frame. The division board, usually

42

called the 'dummy', consists of a board extending to the full depth and width of the brood chamber and is sold with the hive. It is used when the full number of frames is not required, thus excluding the bees from the empty portion of the brood chamber. The brood chamber normally contains at least eleven frames and the dummy board, which is suspended behind them. It may be fitted with a Swiss entrance constructed of metal and screwed in place, the sliding panels being slotted so as to give the ventilation required at any particular time; when fully extended this type of entrance gives an opening of eight and a half inches. The upper section of the fitment can be moved up and down, so that the entrance may be reduced to five-sixteenths of an inch in height or raised to three-quarters of an inch.

The danger of mice or shrews entering the hive must be guarded against, particularly in the depth of winter when the colony is grouped into a cluster or bee nest. Silent and apparently lifeless the colony is an easy prey for these cunning bandits, who creep into the very heart of the group devouring them and building their own nests in the hive; the Swiss entrance forms a secure barrier against such vermin.

The lift or raiser is the portion of the hive which fits on top of the brood chamber enabling the beekeeper to place crates of sections, or frames of comb, often called supers, over the brood chamber, so that the bees can fill them with surplus stores of honey. In winter the lift should be inverted and slipped down over the brood chamber, thus giving additional protection. A second lift can be fitted over the first when it is necessary to place more than three crates of sections or two supers on the hive.

Lastly comes the roof which should be flat, sloping from front to back and covered with zinc. A flat roof can be used for hiving swarms in an emergency by placing it so that it slopes up to the alighting board. In winter a heavy weight can be placed on the roof in order to secure the hive in heavy gales.

QUEEN EXCLUDER—Is a sheet of perforated zinc with oblong holes through which the bees can pass, but which excludes the queen and the drones. It is used to confine the queen to the brood chamber, and is much more reliable than the older type wire excluder. The modern zinc excluder has openings larger and closer together than the older pattern, thus permitting the workers to pass through more freely.

Although the queen excluder will prevent the queen from passing through to sections or frames used for storing honey, it is not safe to put a super of brood frames above the existing brood nest with a separate queen in it. Queens within striking distance of each other will fight to the death, thus upsetting the normal routine of the hive.

THE SKEP—The term 'skep' applies to box hives and straw skeps, the last meaning 'basket' in old English.

In the early days of the skep bees were left severely alone until the end of the honey flow. They swarmed of course in the meantime, and in this way produced new colonies, but swarming only occurred because the queen had not enough room to lay her eggs and the skep had become too crowded for comfort. With the departure of a large part of the colony collection of nectar came to a standstill for a time, mostly at the height of the honey flow, as it does now in a moveable comb hive which is mismanaged. Then as autumn drew near the skeps were weighed and the heaviest chosen for sacrifice, and the owner, fearing being stung, smothered the bees in them in the crudest manner with sulphur fumes. The combs were then torn away from the sides of the skep and the honey allowed to drain into large dishes.

Later an ingenious beekeeper conceived the idea of making a hole in the top of the skep and putting a small container over it, which the bees quickly utilised as a store for honey; thus we have the forerunner of our honey crates and supers.

In some of the old ancestral homes of England I have seen a recess built into the walls of the garden, forming a shelf about eight feet wide and eighteen inches deep, and just high enough to accommodate a number of skeps. An iron bar on one side swung into a socket on the other and was fastened with a padlock to prevent the skeps being stolen; so much did the owners prize the honey and wax their bees produced.

In the seventeenth century Charles Butler spent much time among his bees, and considering the limitations of the skep for observation had a very accurate idea of much that went on under its dome. He had an ingenious way of doubling his skeps by turning one upside down, resting it in a hole in the ground scooped out to hold it, and putting another on top, their open ends fastened securely with pointed sticks and a strong metal band to keep out draughts, but leaving a hole for the bees to come and go. Knowing that in thus uniting two colonies only one queen would be allowed to carry on he left them to settle their differences after making the skeps quite secure. This plan ensured a good surplus of honey for him.

Even with the modern hive and its fittings, the skep still remains indispensible in providing a temporary shelter for a swarm until the beekeeper is ready to hive it.

OBSERVATION HIVE—The glass observation hive was invented by Réaumur in 1830 in an endeavour to solve the problem which tormented him as to how the bees made wax, but in this he was unsuccessful as his efforts only led him back to where he started. Since then there has been little development in this type of hive.

These hives are most useful for the instruction of young people in elementary beekeeping, as they can be kept in the home and the colony studied. They are usually placed on a shelf on a level with a window sill so that the entrance fits under the window sash, the bees passing in and out and taking no notice of anyone in the room.

This type of hive has its limitations, however, as it is impossible to see what transpires in the cells behind the cappings and overcrowding occurs in a short time. Bees winter nicely in these little hives provided the temperature does not reach freezing point or stay below 40° Fahrenheit for a long period, but it is essential to keep bees in observation hives well supplied with food.

FEEDERS—Feeding is practised for two reasons, firstly to prevent starvation and secondly to stimulate brood rearing at times of the year when no honey is coming in from natural sources, and all the honey harvest has been taken from the hive. It is poor economy not to leave some honey in the hive as natural stores go further pound for pound than sugar syrup. There are, of course, times when bees must be fed, and if no honey is available the best food is ordinary white granulated cane sugar syrup; experience has shown that brown sugar is not so satisfactory. Sugar syrup when capped over makes excellent winter stores, but for breeding it is not by any means the equal of honey which contains food elements not present in sugar. No attempt should, therefore, ever be made to extract honey from the brood nest. For this reason honey, or honey diluted with syrup, should be fed when necessary in order to stimulate brood rearing. When all winter feeding has been completed, and before the hives are packed up for winter, a cake of sugar candy one and a half inches thick and six to eight inches square should be placed under the sheet of hessian and over the centre of the brood nest. It must be clearly understood that if this candy is even only slightly scorched it is quite unfit for feeding to bees.

For feeding syrup in spring and autumn a rustless feeder made of aluminium, which is easily cleaned and very efficient, is to be recommended. The following are some useful recipes:—

Syrup for Autumn Feeding
 5 lbs. white cane granulated sugar
 ¾ teaspoonful tartaric acid
 1 qt. water.
Dissolve by heating until all crystals have disappeared, then bring to the boil and simmer for three minutes.

This is for rapid feeding each evening early in September, until the bees have sufficient stores for winter.

Syrup for Spring Feeding

    5 lbs. white cane granulated sugar

    2 qts. water.

Dissolve by heating until all crystals have disappeared, and feed in very small quantities each evening.

It is of great advantage both to the beekeeper and the bees to stir into either of the above mixtures while hot 5 lbs. of honey, making a honey syrup so essential, especially for brood rearing in spring.

Candy

    5 lbs. white granulated sugar

    $\frac{1}{2}$ teaspoonful cream of tartar

    1 pint water.

Stir constantly to prevent burning until all the sugar is dissolved, then bring to the boil and allow to continue for exactly three minutes. Remove from the fire and set in a bath of cold water for a few minutes. Now stir vigorously with a wooden spoon until it whitens like thin gruel then pour into moulds.

An aluminium saucepan is most suitable for making bee candy, which is chiefly used for placing over the brood nest when packing up for the winter.

QUEEN REARING HIVES—The following is a very simple way of rearing young queens from a hive that has shown little inclination to swarm and has proved desirable in every way. Take an ordinary empty hive of eleven frames and insert down the centre a close fitting division board, thus dividing the brood chamber from front to back. Take out a frame which has a queen cell from the hive containing the colony from which it is intended to rear a young queen, making sure that the old queen is not on it or accidentally interfered with in any way. Place the frame on one side of the division board, and taking a similar frame from the same hive also with a queen cell, place it on the other side of the division board, not forgetting to replace these with two empty frames. Now fill up the space left on either side with frames making certain that at least one in each compartment has sealed honey. Then from the parent hive shake about two pounds of bees from the most crowded of the frames over those in the queen rearing hive and, after placing a feeder in position on each side, pack round carefully using two separate covers. Some of the old bees will return to the parent hive, but all the young bees will remain. After two days a little syrup should be fed every evening just before dark.

This operation can only be successful when carried out at the height of the season in full sunshine, and when all hives are at full strength.

When these two young queens have hatched and are mated one can be transferred to a colony which requires a new queen and the hive adjusted to its normal arrangement, leaving the remaining queen to build up the colony to full strength.

The simplest way of introducing a young queen to a hive of bees in the same apiary is to remove the unwanted queen between ten and eleven o'clock on a fine morning, and during the same evening to take the young queen from the queen rearing hive, lightly smearing her with honey and placing her on the alighting board near the entrance of her new home making sure that she goes into the hive.

Full instructions for introducing are usually enclosed with queens purchased through the post.

THE EXCLUDER SCREEN—This consists of a sheet of perforated zinc cut to the external measurements of the hive body, that is $18\frac{1}{2} \times 17\frac{1}{2}$ inches, and framed with quarter inch plaster laths nailed right through with gimp pins. Two cross bars of the same lath spaced equally down the centre are fastened in place with tabs of thin metal in order to prevent sagging, and in the middle of the sheet a Porter bee escape is soldered to the zinc.

This excluder may be used as a super clearer, and can also be left on the hive with the crates or boxes of honey above so that the honey will be maintained at hive temperature until the beekeeper has time to remove it, which is a decided advantage.

COMB—A comb consists of numerous six sided wax cells with one quarter of an inch between their parallel sides in worker cells, and three-eighths of an inch in drone cells, each cell sloping slightly upwards from base to mouth. A mid-rib of wax forms the base of the cells on each side of the comb. Frames fitted with comb foundations from which the worker bees draw out the cells, are provided by the beekeeper to take honey or brood. Queen cells in no way resemble those provided for drones or workers, being much larger, similar to an acorn in shape and pitted over like a thimble. They are about an inch long and usually found on the outer edges of brood combs. Bees also build brace comb which, as the name implies, gives added strength to a frame full of honey. This comb does not consist of a number of individual cells, but is a solid structure which holds the frames firmly in position.

The cell walls of the comb in the brood area are so well insulated that when the bees cluster over it the temperature inside the cell remains constant at 95° to 96° Fahrenheit, with a relative humidity of 70 per cent.

Modern frames for brood combs are so beautifully and exactly machined that there is little trouble in fitting them together. After the joints are driven home a fine nail should be put in each just where they grip the lugs so as to prevent them from pulling apart.

Four holes are bored with a tiny awl through each side of the frame taking care to keep in the exact centre. Through each hole is threaded a strand of wire which reinforces the comb, thus minimising the risk of it falling from the frame in hot weather or when being passed through the extractor. Before wiring, drive in a gimp pin just above the top hole at one end of the frame, and another just below the bottom hole of the same end. Next thread the wire through the holes wrapping the ends around the gimp pins, and with a tap of a hammer drive the latter home. Special soft wire is sold for the purpose by all dealers, but care should be taken to stretch each wire before inserting in the frame.

To make the wire really taut fix a couple of stops on the work bench at just such a distance that the frame is bent inwards when forced between them, then thread in the wire and fasten as described to the gimp pins. When the frame is pulled out from between the stops the sides straighten out, thus pulling the wires as taut as fiddle strings. These four wires equally spaced give perfect support to the foundation.

For fitting the wax sheets, foundation boards are sold by all dealers in bee appliances. To quote Mr. W. Herrod-Hempsall who gives a neat way of getting the edge of the wax sheet inside the slit in the top bar:—

'Drive a couple of nails into the bench about $\frac{3}{4}$ inch apart; these are inserted in the saw-cut of the top bar. When the frame is twisted half round on them, the cut is opened and held there automatically. The heads of the nails should be filed off so that they go into the slit easily, and they should stand up about $\frac{3}{8}$ inch out of the wood. The two top corners of the wax sheet should be cut off, say, $\frac{3}{8}$ inch, and the sheet will then drop easily into the slit. See that it is in squarely, then twist the frame back to its first position and the top bar closes on the wax. A couple of $\frac{3}{4}$ inch oval brads driven through the top bar locks the grip, and the next job is to embed the wires completely in the wax.'

For years I have used the now much despised spur embedder for this purpose and have never had a faulty piece of work; the secret is to keep the embedder submerged in water which is not too hot. An electric embedder is now commonly used for this purpose.

When brood combs become dark it is time to put them in the solar wax extractor as the cells become clogged up with all the discarded silken robes of the emerging nymphs. Bees dislike old combs and their presence is a fruitful cause of swarming, and as they have the urge to build fresh cells in spring that is a good time to remove old combs and give fresh brood foundation.

Sometimes we may find on opening a hive small groups of immature bees looking like silent corpses, with their little white heads in rows just about on a level with the surface of the comb. This is disturbing to a beginner, who imagines that something serious is wrong, and who should in cases of doubt call on expert advice. Personally I have only observed this condition on two occasions, both during very warm weather when the bees for one reason or another decided to leave uncovered a portion of the brood.

SUPER CLEARER—This is a close fitting frame of wood placed under the crate or super so as to make it impossible for any bee above the super clearer to pass out anywhere other than through a Porter bee escape, which is fitted to the centre of the frame. Through this trap bees can pass down into the brood nest, but cannot return, and the latest improvements allow as many as eight bees to pass down at a time. When in use the super clearer should be placed with the round entry hole upwards under the crate or super to be cleared.

SMOKER—A smoker is an essential item of equipment and must be kept in the best of order at all times. Brown corrugated paper makes excellent fuel.

VEIL—A veil should be worn on every occasion while handling bees in order to protect the face. It should be made of fine black net to give ninety per cent. vision, and must be roomy enough to give perfect freedom of movement. Alternatively a woven wire mesh veil can be used, and is cooler to wear than cotton net.

HANDY BOX—It is a good plan to have a handy box for carrying appliances, consisting of a plain wooden box 18 inches long, 9 to 12 inches wide and about 10 inches deep; it should be fitted with a hinged lid and handle for easy carrying.

In this box can be kept a flat knife, screwdriver, chisel, flat scraper, turkey feather, tin of vaseline, box with subduing cloth and stick for same, and smoker; in fact anything the bee-keeper might require at a moment's notice.

THE SOLAR WAX EXTRACTOR—No apiary should be without a solar wax extractor. Beekeepers are handy men as a rule and this can easily be made at home. It consists of a stout box with double glazed lid made to slope much in the form of a cucumber frame. A grooved sheet of tin or zinc rests on ledges just below the lid with a vertical piece of wire gauze for straining the wax from the dross, so placed that it can be easily removed for cleaning. This enables the dross to be held back while the wax runs freely into the container below.

The advantage of this simple contrivance is that it may be placed in a sunny position and is always ready for all odd scraps of comb and wax collected from time to time. The wax melts with the heat of the sun coming through the double glass and runs down the slope into the tray beneath, leaving all the pupa skins and refuse on the wire gauze. The parts can easily be removed for cleaning.

CARBOLIC CLOTH—A carbolic or subduing cloth consists of a square of muslin with a deep hem on one side, wide enough to allow the insertion of a light cane about half-an-inch in diameter. The solution used with the cloth is made up of one ounce Calverts No. 5 carbolic acid to one pint of water. The bottle containing this solution should be labelled POISON.

The first time the muslin is used steep it thoroughly in water, squeeze out and heavily damp with the solution. This cloth so treated may be kept tightly covered in the handy box referred to above. When required, the turkey feather should also be put through the solution.

THE BEE GARDEN, BUCKFAST ABBEY (*page 53*).

A. Worth

THE BEE WALK IN EARLY SUMMER (*page 53*)   *A. A. Lisney*

4

THE APIARY AND THE
BEE GARDEN

*'And the Lord God planted a garden eastward, in Eden;
and there he put the man whom he had formed.'*

GENESIS CHAP. 2, VERSE 8.

THE monks were the first horticulturalists, and orchards and gardens grew around the homes of learning centred at the monasteries. In leisure hours they tended their bees and made their vintage, contributing richly to the history of beekeeping and the skill of making fragrant and sparkling beverages.

In the reign of Henry VIII the gardens of Nonsuch were particularly splendid, especially the fountains, and Spencer is eloquent in his description of these 'silver floods'. About the same period Hampton Court was laid out by Wolsey.

The garden of the Countess of Bedford is immortalised by the poet Donne, with its stone steps leading to extensive parterres and a cloister facing south and covered with vines. These, with the ivyed balustrades and 'walls mellowed into harmony with time,' went to make up gardens that suited and encouraged the meditative temper of our ancestors.

The English garden of the sixteenth century was on the Italian style; the solemn terrace, sloping lawn, sculptured trees and fountains, all reflections of Pliny's Tusculan villa.

Pope had a charming garden of a few acres all his own work, and by degrees the influence of taste began to spread; the Italian garden with its splendid terraces succumbed to Dutch influence introduced by William III, and these in turn gradually gave way to fresh ideas, though remnants of early styles have persisted.

D
53

Cowley wrote to Evelyn, 'I never had any other desire so like to covetousness as that one which I have always had . . . . that I might be the master at last of a small house and garden, and there dedicate the remainder of my life only to the culture of flowers and the study of nature.' In this he echoes the hopes of many in finding an Eden of their own.

> I sometimes see in twilight hours
> The garden of my dreams,
> Where all the scented shrubs and flowers
> Are fed by rippling streams,
> And mossy paths with sheltering trees
> Lead on to seats of restful ease.
>
> The air pulsates with drowsy hum
> Of laden bees returning home,
> The sweet moan of brooding dove
> With scattered song of birds we love.
> In nature, next to God, we find
> That peace which soothes a troubled mind.
>
> All down the ages man has striven
> To make on earth a little Eden,
> Unstinting toil and loving care
> Spent lavishly to make it fair.
> It burns within him since he trod
> The Garden where he walked with God.
>
> A. F. E. L.

One charm of gardens peculiar to these islands is the freshness and beauty of the turf, and the quiet courts of our colleges and universities present the loveliest stretches of grass which can be found anywhere; there is a peace and serenity in their velvet lawns.

In all these lovely gardens, of which I have only mentioned a few, the bee held pride of place; arched recesses built into deep walls, sometimes called bee garths, held the skeps humming with life. Fountains were associated with the bee as Aristotle, Cato, Pliny and Theocritus took much pleasure in watching 'the tawny coated bees fly humming round the fountains,' not for one moment realising that they came for water instead of, as they thought, joining the silver spray in an airy dance.

It is the ambition of every beekeeper to plan a garden with a suitable place for his hives and flowers for the bees. In the smallest garden there is scope for individual taste in the beauty of blended colour, and what a living picture one can paint with the flowers beloved of the bee! A small orchard of choice fruit trees carpeted with that most wonderful of all bee plants *Limnanthes Douglasii* makes a picture never forgotten, and the perfume is delightful.

A gravel or cinder path should be chosen for the hives with a sheltering wall behind, or a fence of sweetbriar and *Clematis rubens* with perhaps a lavender hedge. The hives should face east or south-east to catch the morning sun, and if they are partly in the shade of an orchard so much the better. Herbaceous borders edged with arabis and aubretia in variety with clumps of crocuses and snowdrops should be within easy reach, as all these are a rich source of nectar and pollen in springtime when the bees are busy building up their colonies.

Other flowers with which the beekeeper should stock his garden are *anchusa*, bergamot, *chionodoxa*, *Daphne mezereum*, forget-me-not, grape hyacinths, hyssop, hyacinth, melilot, marjoram, mignonette, michaelmas daisies, *Nepeta mussini*, *N. grandiflora*, *Phacelia viscida*, sage, tulips, thyme and wallflowers. All these skilfully arranged with the owner's own choice of many others go to make a lovely picture throughout the year, providing the bees with nectar and himself with aesthetic enjoyment.

Of shrubs also there are many providing a valuable source of nectar. Cotoneaster heads the list and it must have been a bush of this species outside Cowper's study window which attracted the hundreds of bees that made such delightful music for the poet. All the varieties of *berberis*, lavender, mallows, rosemary, *viburnam* and willows are very attractive to mention only a few.

But above all it is the fruit trees, hawthorn, sycamore and lime that yield nectar in abundance. It is interesting to note that in this country white clover is, next to the lime, one of our most fruitful sources of honey and has a particularly delicious flavour.

Heather honey is popular, but is only within reach of those who live in districts where there are many acres of this plant.

Nectar and Honey

'Honey,' said Aristotle, 'falls from the air chiefly at the time when the stars are shining, and when the rainbow rests on the earth.' Pliny said, 'This substance is engendered by the air, above all when the constellations are rising and more especially when Sirius is shining.'

Nectar is the spirit of the flower; in it is condensed the nutritious sap, delicate fragrance and subtle essences. It is by the labours of the bee that flowers are able to mate with one another, to renew their energies and to perpetuate themselves. The bees give far more to the flowers than they receive from them; it is a service the importance of which cannot be over-estimated for on it hangs the life and health of mankind.

The fruit and seed of the flower come into being when certain conditions have been fulfilled. The flowers must be fully opened and ready for fertilization, and the sun must not be too strong nor the air too dry. It might seem just a chance that all these conditions could arise at the time of necessity. When the precious drop is distilled in the nectaries of the flowers in the gardens, orchards and fields and they are ready for fertilization, they send fragrant waves on the air far and wide to summon the bees. Advantage must be taken immediately of the right conditions as time is fleeting when the mysterious call is sent forth, borne on the soft waves of air currents for miles around. It reaches the hives instantly in the same way as messages travel through the air reaching us on the radio as the morning news.

We who live so much with our beloved bees have become attuned to these fragrant waves in a much lesser degree and can pick up this call from the flowers. We feel a magnetic quality and a pulsating sweetness in the air which is over-powering, and hurry up to the hives to see if we were correct. There are the bees pouring out in a golden stream, when a short time before there was not much activity.

The bees are equipped to receive these messages with their antennae which contain olfactory cavities of the most exquisite delicacy. All the bees in the hive get the message at the same time, just as I believe all the bees in a swarm receive as one the fragrant scent from the scouts, wafted on the air from the tiny crescent which appears on the last segment of the abdomen. The olfactory nerves of man are not sensitive to such delicate scents.

Very early in the morning, long before the sun has dried the dew from the grass, the lime trees send out the call and we hurry to the trees to hear the delightful murmurous chant from thousands of bees drinking the intoxicating nectar. Fully laden they return and drop heavily on the hive, the alighting board and the grass, to rest before entering.

Lime trees are one of the richest sources of nectar, yielding a delicious golden-green honey which has a faintly narcotic effect on bees, more especially the bumble bee who becomes quite intoxicated from drinking too heavily. Clover yields a delicious pale golden honey of the finest quality, but hot humid days are necessary for the harvesting of this crop.

It has been asserted that, as the result of an experiment, a bee which has discovered a source of nectar returns to the hive and informs the others by breaking into a lively jig on the honey comb. In the experiment referred to a bowl of syrup was placed at a certain distance from the hive, and shortly after the first bee arrived to partake it started dancing; very soon it was approached by perhaps two others who joined in the dance and within a few minutes they were all out imbibing. These bees in turn 'homed' to the hive and danced so that more workers were encouraged to seek the syrup. Each day the bowl of syrup was moved further away from the hive until the distance was over a hundred yards, then the pioneer bee did a difference dance and capered along a broad figure-of-eight course wagging its tail each time it did the straight steps across the line of dance. The nearer the syrup was to the hive the faster the bee danced, and when it was a hundred yards away it did forty tail wagging runs down the straight every minute; when the syrup was a mile away the number of runs was eighteen. I find it not easy to accept this theory, but I shall return to the subject later.

A populous colony will in a good year collect at least one hundredweight of honey as well as rearing countless brood; the precious harvest must be gathered even at the expense of other duties. This love of honey in the bee is absolute in its power, and demands a patient and persistent effort which exhausts the bee and shortens its life. The foraging bees do not pass on the nectar to any others when entering the hive, but themselves spread it over as many cells as possible.

I am often asked why bees gather so much more honey than the consumption of the hive demands. The answer is given by Lord Lubbock who says, 'I believe that this great avidity for honey must be attributed rather to their love of the common welfare than to the selfish gratification of their individual appetites.'

The honey gatherers go to the humblest flower on the slopes of the hills and in the roadside ditches, the bees never losing their way in barren solitudes, never wasting time, but flying straight to that drop of moisture which oozes sparingly. This nectar has a very different consistency from the matured honey in the combs, and being diluted with water it must be concentrated by evaporation in order that it may keep for long periods without fermentation. On returning from the harvest the bees distribute their nectar in shallow layers over as many combs and as many cells as possible. However abundant the reserve stores rushed into the hive, the surplus will never seem enough to the indefatigable worker.

Nectar must be reduced to about a quarter in weight, and the bees themselves are obliged to expel excess water from that which they collect. A certain amount is eliminated while they are on the wing, but concentration by evaporation is solely the result of the unaided effort of the harvesters acting as ventilators. Through the long dark hours when all the world is hushed, the valiant workers perform the prodigious feat of eliminating from the hive about two pounds of water in a single night, and I have seen it dripping from the alighting board which was covered with moisture. By what exquisite sense do bees know when the honey has acquired the requisite density we do not know.

It has been discovered that honey contains quite a perceptible trace of formic acid, the smallest quantity of which will prevent or arrest any tendency to fermentation. It is said that at the moment of sealing each cell the worker bee instils a minute quantity of fluid from the sting sac which contains this acid. This also has the effect of accentuating the mellow sweetness of the honey and the lingering fragrance of the blossoms which it contains, thus rendering it something more than a delicious food and remedy for many ills. When each cell is full to overflowing it is sealed over with a mixture of wax and propolis, but if the bees are of the black variety the cell is capped before it is quite full, giving it an ivory whiteness that is not achieved by the Italian bee which places the capping directly on the honey.

This substance, limpid and fragrant, was in the earliest times imagined to be a food for the gods who were not believed to be immortal unless they feasted on a heavenly fare of honey; the ambrosia of the gods. Why they should need food when they were considered to be immortal is an unanswered question. Sappho wrote—

> 'The goblets rich were with ambrosia crowned
> Which Hermes bore to all the gods around.'

The sum of all that was richest in nature, honey became the universal panacea. It was the miraculous honey which the Argonauts poured forth in generous libations on the waves of the favouring sea, while the Romans mixed it with their wine. The offering of honey libations to the dead was a very ancient custom and one of the oldest religious rites which continued for thousands of years. Sometimes the dead were preserved in honey.

The bees which lived in clefts of rocks or in caves in the East were considered as being intimately connected with, if not embodying, the souls of the dead; the clefts and caves being thought to be entrances to the underground world of dead spirits. Virgil may have been thinking of the connection between the souls of the dead and bees when he described Aeneas beside the waters of Lethe, and compared the spirits who were drinking the water to bees swarming in the meadows beneath the cloudless autumn sun. Euripides makes Helen send Hermione to the grave of Clytemnestra with offerings with these words—

> 'Take these offerings in thy hands. Soon as thou
> reachest Clytemnestra's tomb, pour mingled
> streams of honey, milk, and wine.'

Bees were also associated with the worship of such gods as Demeter, Rhea, Zeus, and the priestesses of Demeter were called Melissae. Pan and Priapus, often mentioned in latin books on beekeeping, were believed to be protectors of gardens and bees. On a statue of Pan was written—

> 'Having left the slopes of Maenalus I abide here
> to guard the hives, on the watch for him who
> steals the bees.'

The belief that honey came down from the skies to be collected by the bees was very general many hundreds of years ago, and as late as the middle ages it was believed that if bees were not informed of the death of their master they flew up into the sky to seek him there. It was customary to tell bees of occurrences which happened in their owner's family especially of a death as they, being the winged messengers of the gods, carried the news to spirit land of the speedy arrival of a newcomer.

It is not surprising then that bees were held in the greatest reverence by man in ancient times; for him honey was everything. For many thousands of years it was the only sweetening agent that existed or which man can have known; the only substance capable of lending to the most insipid foods the relish of an almost magical taste. Man blessed the bees and held them in the highest esteem for bestowing on him such a rare and incomparable delicacy.

We have other sugars now, and even if those we manufacture are not as pure and have no fragrance, at least they are made in great abundance and with ease. If honey has lost its importance, not so the bee.

The gathering of nectar is only one side of the whole mystery surrounding the honey bee, and not as marvellous as the benefits bestowed on mankind through the fertilization of the flowers when they surrender to the bee their essences stored in the nectar. Flowers are dependent on the bee which, in exchange for a drop of fragrant liquid, assures them of fertilization.

THE HONEY FLOW

The honey flow is an expression well known to beekeepers to describe conditions when bees can be observed flying swiftly from the hives and returning obviously heavily laden. They fall on the alighting board or on and around the hive which they enter after resting awhile; there are thousands of them, yet half an hour or so previously they did not appear at all busy. When the air is warm and moist, and pulsates with the incomparable waves of fragrance from the flowers, the industry of the bees rises noticeably in tempo thus constituting a honey flow. Many honey flows may occur during the season.

COMB OR SECTION HONEY

When the bees have made good progress with the first honey crate placed on a hive after the spring clean, it should be raised and a second one placed in position underneath making sure that the lower edges, which rest on the brood chamber, have been well rubbed over with vaseline. As soon as the second honey crate is nearly full a third can be placed underneath it and so on. In a good season I have had up to six honey crates on one hive, and am never in a hurry to remove the full crates as honey keeps better in the warmth of the hive until it is convenient to remove it for use or extraction. The best way to remove crates of honey is by using a super clearer equipped with an up-to-date bee escape, the under surface of the super clearer being always well smeared with vaseline before it is placed in position.

It should never be forgotten that all work in connection with bees must be undertaken in fine weather when they are flying freely; in no circumstances should the hive be interfered with in rainy or threatening weather, or after dark.

In storing comb or section honey throughout the winter months it is necessary to keep it free from damp as well as cold. The best method of preventing these dangers is, of course, to store the honey in a place where neither damp nor extreme cold can reach it, and nothing is better than a shelf close to the ceiling in a warm kitchen.

Honey freezes at a relatively high temperature so that, whether extracted or left in the comb, it is liable to cloud and partially solidify at the approach of cold weather. At the same time, however, there are kinds of honey which taken on the same solid form after being removed from the hive, even in warm weather.

This tendency to solidify is in fact one of the best indications of the purity of honey, and frequently liquid honey distributed commercially has been diluted with other substances and does not solidify even in extreme cold.

A large biscuit tin is ideal for storing section honey; it should be lined with grease-proof paper before the sections are put in, and stored in this way in a consistently warm atmosphere the honey will keep indefinitely. A biscuit tin holds sixteen sections.

EXTRACTED HONEY

A pair of good uncapping knives will be needed, and a honey storage tank consisting of a tin container holding one hundredweight of honey. It has a strainer on top covered with a lid, and a tap at the bottom to draw off the honey when necessary.

A good centrifugal extracting machine is a wise investment for anyone who has more than four hives; the small size will extract two standard frames, and the radial pattern twenty or more combs at a time. A tap is fitted at the bottom for drawing off the extracted honey.

The honey flows from the extracting machine, passing through the mesh of the strainer into the storage tank where it can be kept until drawn off into bottles or jars. As honey usually solidifies very rapidly when taken from the warmth of the hive it should be extracted as soon as possible, and must not be left in either the extractor or storage tank for longer than necessary.

The knives employed to uncap the combs must not only be sharp but heated before use, as this helps to cut through the wax of the capping. This is done by allowing them to stand in a deep can three parts full of water kept simmering on the cooker. The knives can be used and returned to the can alternately throughout the work, after each has been passed over a sharp edge to clean off any adhering wax.

In uncapping, the lower end of the frame must rest on something solid and the top lug grasped firmly in the left hand, with the comb tilted slightly towards the knife so that the cappings fall clear on to a tray as they are cut off. The cut is started at the bottom of the comb, and with a steady sawing sweep continued upwards so that the caps fall away in a solid sheet. Both sides of the comb must be uncapped before the frame is placed in the extractor, and after a little practice proficiency in this operation is gained.

The exact speed at which the machine should be revolved is a matter of judgment and can be acquired after a little practice. The cages should rotate only just fast enough to empty the comb, any excess of this speed being liable to damage the combs against the wire mesh sides of the cages, or even pull the combs out of the frames altogether.

When at the end of the season all combs have been cleaned by the bees they must be stored away in a cool place. A teaspoonful of P.D.B. (para di-chlorbenzene) crystals placed with each box of frames wrapped in newspaper will greatly reduce wax moth trouble. These combs can be used repeatedly for many years, assuming of course that they have been correctly wired as otherwise distortion will appear.

It is of the utmost importance that all appliances used for extracting should be scrupulously clean. Before and after use the mechanism should be run for a while in water as near to boiling point as possible, after which it is dried thoroughly and all moving parts oiled with machine oil. I have had the outside of my extracting machine and storage tank coated with aluminium paint so as to prevent corrosion.

POLLEN

Pollen constitutes the protein food of the honey bee, and is the male element necessary for the fertilization of the female ovule in flowering plants. It is to the necessity for bringing together these elements that we owe all the beauty of colouring, design and scent.

A flower consists of the ovary, which is the pouch containing the seeds or ovules; the style which is the slender rod arising from it; and the stigma with its glutinous coating. All these form the female portion of the flower. The male part consists of the anthers, which are bags of pollen mounted on thin stalks known as stamens and are usually found clustering round the central style. A grass or an oak tree has flowers just as much as a canterbury bell or a primrose.

There are two large groups of plant life; the Anemophiles, from the Greek *anemos* meaning the wind, which includes most of our trees, grasses, and sedges producing their pollen at a convenient season. The pollen is well exposed so as to be easily carried away and blown into the meshes of the feathery and sticky stigmas of their kith and kin; it is smooth and flowery and adapted in every way for air transport.

The other group, the Entomophiles, from the Greek *entomos* meaning an insect, which is distinguished by the possession of various scents and coloured petals. Here we find a definite partnership with insects carried to the utmost limit, and plants in this group deprived of visits from certain insects never set their seeds at all.

What then is there about pollen which makes it of so much importance to the bee? We see it coming into the hive in the form of compact balls and cakes of many colours carried on the corbiculae of the bee's third pair of legs. Sometimes the bee is completely smothered in the dust of some particular variety of plant, and from the colour and texture of the pollen it is often possible to tell which plant the bee has visited; for instance the mallow envelopes it with a pale blue veil.

The importance of supplies of fresh pollen in spring for the building up of the young brood cannot be exaggerated. This product of plants is rich in protein which is an absolute necessity for the growth of all animal bodies, human and insect; there is nothing that can take the place of protein in the development of the larva.

The bee enters the hive with a veil of coloured dust over it, and with little balls of pollen held in the small triangular depressions by long curved bristles that border the tibia of each hind leg. This is the result of visits to the heart of the flowers lasting up to perhaps one hour. The bee always visits the same kind of flower on each journey so that these tiny pellets are all of the same shade. For example, if the bee visits the hazel catkins it heeds no other source no matter how tempting; or if the gorse is visited only pollen from this flower is collected until the bee returns to the hive.

The bee brushes the anthers with sweeps of its legs to collect the pollen which is, with the assistance of the stiff bristles on the joints of the forelegs, passed to the middle legs. These in turn pack it into the baskets of the hind tibia and the bee flies off to another flower. Finally the bee brushes the last particle of gold dust from its thorax and abdomen, adds it to that already in the well packed baskets and flies back to the hive. It is well worth while watching all this taking place on flowers and shrubs which yield a rich supply of pollen.

It is also an interesting study to watch a bee laden with pollen enter the hive. Tired? Not a bit! It swaggers around as proudly as a bee could look and is in no hurry to divest itself of the precious burden, running all over the frame. Just when one's patience is becoming exhausted it chooses a cell, and clinging to the edge plunges its hind legs into it. With the middle pair of legs it detaches the two pellets so that they fall into the cell, and by means of its head rams them hard home. That accomplished, the bee brushes itself and stretches out its antennae with the forelegs in order to clean them with that most familiar of all gestures every bee repeats hundreds of times daily. Though the bee is careful, when collecting pollen in the fields, not to mix the pollen of one species of flower with that of an entirely different one, pollen from different plants is placed in the same cell resulting in a mosaic of colour.

As it is a matter of keen interest to beekeepers to know from what scources come the various kinds of pollen, the colour of a few of the more usual may be mentioned:—

Dandelion and Broom ...	Deep orange
Gooseberry	Dark green
Sycamore	Greenish yellow
Apple	Very pale yellow
Crocus	Orange
Charlock	Pale yellow
White clover	A sober brown
Red clover	Dark brown
Sainfoin	Brown
Pear	Crimson
Poppy	Dark purple
Blackberry	Greenish white
Raspberry	Similar to apple
Sallows	Greenish yellow
Gorse	Yellow to dull orange.

PROPOLIS

Propolis is of Greek derivation, *pro* meaning before and *polis* a city, indicating its use in partially closing the entrance or gateway to the hive. It is gathered by the bees from various trees and is of a gummy consistency, dark reddish brown in colour with the odour of pine. Bees gather propolis with their mandibles, and it is packed and carried in the pollen baskets in the same way as pollen. They do not pack propolis in cells as they do pollen, but as it is very soft when freshly gathered they spread it with their tongues as a varnish over the inner walls and floor board of the hive, where it acts as a preservative and damp proof coating.

The bees begin collecting propolis in any quantity after the last of the honey flow, and as autumn draws near they pour in and out of the hives in a stream feverishly gathering this sticky substance for the insulation of their homes during the winter months.

The quantity of propolis brought in depends partly on the

race of bee and partly on the locality. Caucasian bees use
it much more freely than Italians, and very often the entrance
to their hive is almost closed with pillars and walls between
which is left barely enough space for a single bee to come and
go. In the East, buds of the balsam poplar are believed to be
one of the chief sources of propolis.

QUEEN CELLS

It is important to choose a particularly good morning if one
wishes to examine the brood chamber for queen cells, and this
should preferably be done when there is a honey flow. Bees
resent interference in bad weather or during a dearth of
nectar.

In a hive where a young queen had emerged, been mated
and was laying, I have known the bees, with lamentable lack
of intelligence, ruthlessly kill her after I had opened and
examined the hive. True, the weather was unsettled, but that
was not a sound reason for depriving the hive of its queen and
bringing about the extinction of the colony. It is only natural
that when a new queen has been introduced successfully and
is laying, one is anxious to see how things are progressing.
It is, however, advisable to leave the bees alone for at least a
week after introducing a new queen, when they should have
completely settled down.

It is not usual to find black bees attempting to supersede a
queen as they go in more for laying workers, while on many
occasions I have known Italian bees supersede the queen and
have seen both her and the young virgin walking over the
combs in perfect harmony, only to find later that the old
queen has disappeared. It is very helpful to mark queens and
so distinguish the old from the young with certainty.

I strongly suspect, though I have never been able to prove
it to my entire satisfaction, that the queen does not lay the
eggs in queen cells, but that these are transferred from
ordinary cells by the workers. In support of this theory I have

H. Checkley

BEE GARTHS AT PACKWOOD, WARWICKSHIRE, (page 54).
PROPERTY OF THE NATIONAL TRUST

WORKER BEE ON APPLE BLOSSOM (*page 56*)   *R. V. Roberts*

seen queen cells, with eggs in them, attached to honey comb in supers above the queen excluder.

In April one year, while transferring bees from their old hive to a fresh clean one, I saw to my surprise a lovely young queen that had apparently just emerged walking over the combs, the old queen appearing quite unconcerned. It was too early for any drones so I knew that the life of the young virgin would be short. It seemed a risky thing for the bees to produce a young queen at this time, as the queen and princess could have killed each other. Doubtless, however, the bees had no intention of allowing this to happen as the young queen disappeared later.

On one occasion in a hive that had swarmed, I took the old queen away and hived back the swarm leaving two good unsealed queen cells. I went over this hive carefully a week later leaving only one of these cells, which was sealed and much larger. After a few days I looked to see how it was getting on, and noticed that the bees were not fussing over it as they normally do in anticipation of the emergence of the young virgin, soon to become the mother of the hive; they were really quite indifferent though the cell was due to hatch. With a sharp penknife I carefully removed the huge cell and took it away for examination. To my utter astonishment I found that there were two queens in it, apparently perfect in every way but alas both dead. Twins from birth they had been fed together until the ninth day when a worker bee had capped their cell, little knowing the awful tragedy that would be played out in it later converting the cell into a tomb. Scarcely had the cell been capped over their heads when the two embryo queens had set to work to spin their cocoons. Like the silkworm, the larvae are endowed, for only a few hours use, with special glands from which a secretion mixes with their saliva and enables them to weave a filament of some strength. And so the twin larvae spun their cocoons like softly padded robes and light as gossamer. Thirty-six hours

E

did they devote to this work after which they both remained motionless while passing through the mysteries of that wonderful metamorphosis all bees experience, blissfully unconscious of the destiny awaiting them. As soon as the hour of emergence arrived it is possible that they recognised each other as dual claimants for the supremacy of the hive, for both were dead when I found them. It appeared that, impelled by that fierce hatred all rival queens possess, they plunged into each other the lethal weapon nature has provided for no other purpose. This episode may well be unique in the annals of beekeeping.

SWARMING

A colony of bees that is normal and prosperous increases its brood in the spring, and the population grows until all the available brood comb is occupied.

In early spring only brood producing worker bees are reared, but as the colony becomes stronger the rearing of drone brood begins, thus providing male bees in anticipation of swarming. Finally, when the combs are all filled with brood and the brood chamber becomes crowded with emerging and recently emerged young bees, queen cells may be started. When eggs are placed in these partly built queen cells the colony has taken definite steps in preparation for swarming, which takes place eight or nine days later. Swarming, therefore, normally occurs about the time the more advanced queen cells are sealed, but if there is a prolonged period of bad weather swarming may be abandoned for a time or perhaps altogether. If, however, there is a very hot spell a swarm may emerge from the hive before any of the queen cells are sealed, and this usually takes place between ten o'clock in the morning and three o'clock in the afternoon.

In order to explain the departure of a swarm, which deserts the hive in the full height of its prosperity, it is necessary to go back to early spring. If hives are fed on honey syrup from the beginning of March till the end of April, which is the normal time for a spring clean, they should be ready for the first honey crate. This early feeding encourages the hives to build up, and by the end of April there should be an enthusiastic population of young harvesters all ready when the orchards send forth the call for the fertilization of the blossoms. There is nothing at this stage more stimulating to a populous hive than to have a honey crate or super to fill, but from this moment the beekeeper requires to keep a strict watch on all hives in order to anticipate their wants in advance. As soon as the filling of the super is well advanced another should be added underneath, and while the hive is open the brood combs should be examined carefully for queen cells and all these removed with a sharp knife. The time to choose for this operation is between ten o'clock in the morning and noon on a very warm day, when the bees will be out in the fields and the brood nest consequently fairly clear.

From experience one finds that hives properly managed will not swarm during a honey flow, as the laying activities of the queen are suspended at that time in order to enable all the bees to take part in the gathering of the harvest. If, however, the beekeeper has allowed preparations for swarming to mature before a honey flow begins he will have lost much of the harvest as the bees will have become swarm-minded, thus giving the owner considerable trouble.

Italian bees are not so much given to swarming and are much more easily controlled in this respect than the black bees, which swarm on the slightest provocation and require all the wisdom of the beekeeper to keep them within bounds. In forming new stock a young queen from a colony which shows little inclination to swarm should always be chosen, and I have found by experience that, for some reason which I cannot explain, a queen with a very pointed abdomen should be avoided.

It is when a spell of dark cloudy or wet weather interrupts a honey flow that bees make preparations for swarming; perhaps the bees are instinctively impelled to go elsewhere on the chance of doing better. It is important, therefore, to look over the hives and remove all queen cells or take other manipulative measures, when the sun comes out again after a few dark days have interrupted the honey flow. In spite of all that has been said, however, swarming may be provoked by circumstances beyond the immediate control of the beekeeper.

In order to return a swarm to the colony it has just left it is necessary first to go over the hive thoroughly and remove every queen cell, making sure that there is plenty of room for the queen to continue to lay and frames for the bees to store honey; if these precautions are not taken the bees will kill the old queen. The correct time of day to return a swarm is just before sunset.

According to Maeterlink 'It is the spirit of the hive that determines the hour of the great annual sacrifice to the genius of the species—swarming.' What is the spirit of the hive? Is it the instinct which impels the bee to spread abroad its kind?

Preparations for swarming continue for a number of days, and when the time comes to leave the hive the bees are careful to take with them a five days ration of honey and plenty of pollen. This precaution is necessary in case there is a spell of bad weather after they arrive in their fresh quarters.

As the swarming bees pour from the hive in a brown or golden stream according to the species, they circle in the air over and around the home they are leaving. Spreading out in a great network they drop to the ground covering it with a mosaic of insect life, then rising again the air pulsates with the humming song of these happy and for once carefree workers. This continues for some time until suddenly some mysterious signal is given and they draw together over a bush, branch, post, or whatever they elect to cluster on. Sometimes the swarm divides, part beginning to settle on the branch of, say, an apple tree, the rest in the heart of perhaps a gooseberry bush. After a time the queen will join one of the groups when the remaining bees will immediately follow her example thus covering her.

Gradually the wonderful music is stilled as the bees crowd closer and closer together, till they hang from the branch like an enormous and lovely fir cone. They are very still waiting for news from the scouts which left the hive in advance, travelling further afield to find their new home. This may be a hollow in a tree, an empty hive, or an opening in the roof of a house. If the day is warm and the sun shines hotly on them, they do not take long to decide which of the attractive abodes recommended by the scouts is most to be desired; the cluster falls to pieces and the bees rise in the air to follow the scouts over hill and dale in a definite direction till all are lost to sight.

I am inclined to think that the queen may sometimes enjoy the airy dance at swarming time as much as the bees appear to do, though more often this is far from being the case. It has been with much curiosity on this point that I have closely observed many swarms. The queen is almost the last to leave the hive, borne forward on the crest of a great wave of bees. I have, however, on occasions seen her turn and re-enter. When this happens the bees at first appear as if nothing unusual had taken place, the swarm floating and swaying in the air and eventually making efforts to cluster; a restless seething mass with no cohesion. The bees, soon realising that the queen is not among them will return to the hive; but they swarm again at the next opportunity.

I vividly recall an exciting instance of the reluctance of a queen to leave the hive. She was swept along in the golden stream as the bees issued from their home and kept rising above them, repeatedly turning towards the entrance only to be driven away by the advancing tide. Obviously the queen did not mean to go if she could help it, and as she rose in the air for the last time she very cleverly darted back into the hive. By this time there was a swarm of considerable strength in the air and I presumed that she took advantage of the fuss to slay her rivals in the much depleted colony. The swarm returned to the hive and settled down peaceably.

The beekeeper is never anxious to lose a swarm, as it is much too valuable, so he should get busy when he sees the swarm emerging and decide which hive will be its new home. If the beekeeper is too busy to attend to them at once the swarm, when it has settled in a dense cluster, should be sprayed lightly with water through a fine rose and covered up with a sheet.

To remove a swarm from its temporary resting place an empty skep should be held inverted under the cluster, which is then given a quick dexterous shake so that the bees fall into it; at least most of them do, and if the skep is immediately placed face down on the ground below the tree or bush the rest join those already in it. The mouth of the skep should then be covered with fine muslin and tied firmly in position, when it can be taken to a shady place and left on the ground with one side raised up with a stone to allow the bees plenty of air. There they are quite safe till the beekeeper is ready to hive them. It is absolutely essential to cover the mouth of the skep with muslin, as returning scouts will seek out the swarm and lure it away if this precaution is not taken.

On one occasion I sent to the South of France for a swarm of bees, which arrived towards the middle of April in a blizzard and the ground white with snow. I had a hive ready for its reception and though the swarm had travelled by rail and sea such a great distance, it arrived very much alive in spite of the cold; I had had misgivings as the temperature was at freezing point. I carefully opened the oblong box which had a fine wire mesh panel down each side, and after giving the bees a puff of smoke I shook them gently on to a sheet which had been spread over a wooden platform sloping up to the opening of the hive and resting on the alighting board. They were very slow in moving which I presumed was on account of the cold and I became anxious. I tried to coax them with a feather, but no! they appeared to wish to return to the box. I picked up the box and had a good look inside to see what was attracting them, and there in a small cage fixed carefully in one corner was the queen. I let her out among the bees and she immediately scrambled over them and into the hive. Then a little group of bees lined up on the alighting board standing in the familiar attitude facing into the hive, no doubt conveying the joyful news that the queen was entering the portals. Immediately there was a loud humming response from the entire swarm as they moved swiftly towards the entrance.

As mentioned earlier I have every reason to believe that scouts lead a swarm to their fresh abode in the same way, that is by exposing the scent gland on the abdomen, as soon as it is settled which scouts they will follow. By this means the swarm cannot go astray from the prescribed route.

I had an instance of this when a swarm came to my apiary one July. There was a queen rearing hive with some empty frames ready in case it might be wanted. One morning I noticed a few bees flying around it and thought that they were inquisitive workers from my own hives. A few hours later the numbers had increased, but as there was no honey in the frames robbers could be ruled out. The bees seemed to be giving the little hive a thorough examination, examining every joint and crevice, especially under the eaves of the roof. I lifted off the latter and found considerable activity inside, the bees going over the frames and feeling along the walls. So engrossed were they that not the slightest notice was taken of me. I spent all morning watching them full of curiosity, but the idea of a swarm never entered my head.

About two o'clock in the afternoon I was aware of a humming sound in the distance, and looking towards a nearby wood I saw a small black cloud coming towards where I was standing by the hive. Pausing in the air above the hive they began to fall on it like rain, and there on the alighting board was a group of bees lined up facing the entrance each exposing the Nassenoff scent gland. There is no doubt that these were the scouts which led the trusting swarm to so desirable a haven.

I have at present some hives in which the stock is bred from generations of bees showing little tendency to swarm and which have a noticeable strain of Italian in them; one hive has not swarmed during the reign of two successive queens and the present one has passed her second year. I have seen these bees clustering on the outside of the supers which would lead one to believe a swarm was imminent, but on examination it is invariably found that no queen cells are present. I never keep the queen of a hive that shows an over tendency to swarm after skilful treatment has been applied to prevent it.

When a swarm deserts a hive it means the dislocation of working conditions at the very height of the honey flow, but swarming can be prevented by anticipating the wants of the hive so that they have no desire to leave it; and by requeening all hives which have a propensity for swarming from hives that have not.

Brood combs which have lost their freshness and have a lot of drone comb in them should be discarded. I have repeatedly heard it said that swarms work much harder than the old stock they have left and that it does them good to get away; I agree, they work much better when they have frames with fresh foundation and seem to delight in building new cells for their larvae and for storing the precious nectar. The old colony is left with the ageing and perhaps damaged combs, and has no incentive unless given some fresh foundation when a considerable increase in output will be observed. I have seen a nucleus of bees with four frames fill six new ones in a fortnight with honey and brood, and ready for a honey crate or super. If a colony is encouraged to build up its strength and store honey, they will work like a swarm all the time.

It is well worth while having a small queen rearing hive

holding four or five frames. From a hive that is not given to swarming two combs of brood with eggs, as well as all the bees adhering to it, should be removed, making certain that the queen is not there also. These combs should be placed in the queen rearing hive, which should then be filled with empty combs. A certain number of bees will fly back to the parent hive, but sufficient will remain to rear a queen which, when she has begun to lay, can be transferred to whichever hive is to be requeened, first making sure that the unwanted queen has been removed. This is a very simple and sure way of breeding from the best stock.

The active season in my district closes down during the third week of July at the latest, but in any locality all hives should, before the end of this month, be provided with queens which must carry them through the coming winter and following season, and by this time the mating of all young queens must be assured as drones will not be tolerated much longer. All honey for extraction should be removed before the end of the month, and in this way the bees are left to store in the brood nest all nectar gathered, thus giving valuable winter stores.

The bees will keep working up to the end of September, gathering the scanty though precious nectar that abounds in the neighbouring wayside hedges and fields. Ivy, for instance, yields a valuable harvest late in the season, though it is fortunate that this plant flowers so late as the nectar is bitter and would spoil the flavour of stored honey.

There is a story told of how bees came to Ireland. In one of the many monasteries founded by St. David of Wales where the monks kept bees, one of these monks, St. Modomnoc an Irishman, set out to sail to Ireland; he was very fond of bees and a large swarm followed him and settled on the prow of the ship. He returned with them to the monastery and tried to slip away unobserved, but the swarm again followed him. After the bees had followed him a third time he went to St.

David and asked if he might take the bees with him, so St. David blessed both him and the bees and they proceeded to Ireland where they prospered exceedingly.

LAYING WORKERS

If a hive is allowed to become queenless for any length of time and has no means of raising one, laying workers may appear and take over the duties of a queen. How these imperfect females become capable of laying eggs that produce only drones in no longer a mystery.

It is now considered that when the bees realise they are queenless they begin feverishly feeding other bees in the hive with brood food in an effort to produce a queen, and it is this special feeding which seems to cause a certain amount of activity in undeveloped organs, resulting in what is known as a laying worker. In appearance this freak does not differ from other worker bees, so it is quite impossible to find her in the hive; her eggs are unfertilized, producing only drones, and are laid indiscriminately all over the hive in any cell that is convenient. Sometimes four or five eggs will be laid together in one cell. Drones appear in great numbers after a time, rather undersized but otherwise perfect.

In such circumstances it is as if the bees had an idea things were not just quite right, for they build queen cells here and there over those which contain drone eggs; and yet the poor deluded workers cannot be persuaded to accept a queen if an attempt is made to introduce one to the hive in the presence of a laying worker. Presumably they see eggs and larvae in the cells and imagine all is well. A distinct lack of intelligence is thus exhibited as the workers are normally able to tell the difference between the sexes, and they must be aware that males consume but do not produce. Normally this ability to discriminate is shown clearly enough for, if the food supply runs short, it is the males, both larvae and adults, which are the first to be sacrificed, and at the end of the season they do not hesitate to carry out the annual massacre of the drones.

Why then do they go on producing the very creatures that will cause the extinction of the colony, and resolutely refuse to accept a queen which would at once proceed to do her best to set matters right? These are riddles we cannot easily answer. Unless we come to the rescue of a colony afflicted with laying workers it is doomed to certain extinction.

If a hive from any cause becomes queenless, it is essential to make sure that it has worker eggs, or unsealed larvae of the proper age to produce a queen. Although larvae are fed on a highly nutritious diet for the first three days, it is only for the development of a queen that this rich food is continued. It follows, therefore, that larvae over five days old cannot be fed to produce a queen, and the simplest way of dealing with a colony which has become queenless and has developed laying workers, is to unite it to the weakest hive with a queen in the apiary, providing both colonies are healthy.

Finding on one occasion that a hive of black bees had become queenless I sent to Italy for a queen, but by the time she arrived this hive had produced laying workers. I placed the cage containing the new queen in the hive and a week later looked in to see how things were. The queen was still in the cage looking very scared, and although all the soft candy which was in with her was gone she had not made good her exit into the colony. I plugged up the hole again with soft candy and left her, but when I looked into the hive a few days later she was still in the cage and the candy gone as before. I induced her to come out, and as the bees did not attack her though they looked hostile, I left her among them. A few days later she was outside the hive looking as if life was not worth living, so picking her up and dipping her in honey in the hope that by the time the bees had cleaned her she would be acceptable to them, I again put her back in the hive. About a fortnight later I opened the hive; the new queen had been laying and there was sealed brood, but she was the ghost of her former self. As I was looking at her one of the bees rushed forward, caught and shook her; another followed up the attack and apparently this had been going on for some time as all the hair was entirely worn off her body. In the end she gave up the unequal struggle and the bees were united to another colony.

Another reason which may induce a colony to produce laying workers is an accident to a young queen. When a colony sends out a swarm which is hived in a new home there are a number of queen cells on the frames in the hive they have left, and if the remaining bees decide that the first queen emerging will be the future mother of the colony she is allowed to destroy all the other queens. This she will do by stinging them, either through the walls of their cells or after they emerge; in fact she requires no encouragement to proceed with this slaughter.

The new queen has yet to take her maiden flight, a hazard for which she is perfectly equipped in every way and for which she prepares with scrupulous care. Probing out the meteorological conditions of the day and hour with marvellous accuracy, a virgin queen is rarely lost though there are dangers which she cannot foresee; there is, for instance, the possibility of her being snapped up by a swallow or swift. If the queen never returns to the hive it is left desolate; by this time there are no larvae left which can be specially fed to take her place, and it may be due to a rare accident like this that laying workers appear. Italian bees do not encourage laying workers though they may be queenless for some time; not so the black bee which, very soon after losing a queen, produce this extraordinary product of nature.

Winter Management

When September is drawing to a close it is time that feeding of the stock for winter is completed as quickly as possible, giving to each colony from twenty-five to thirty pounds of honey syrup, if there is not already sufficient honey in the hive. Feeding should be completed by the middle of October at the latest.

Opinions vary as to the best covering to have next the brood frames during the cold winter months, but after years of experience I do not think that this is a matter of vital importance. It is, however, essential to ensure that all hives be at top strength and have adequate provisions for the winter. For upper quilts any old material will serve, such as a blanket cut into pieces, each twenty inches square, and packed in six to eight thicknesses over a square of strong hessian, which in turn covers the brood frames. Every hive needs a plentiful supply of these woollen coverings, and one should be able to draw on a fresh stock of dry squares at short notice to replace any that have got damp. Some use chaff cushions, but they are not as serviceable as squares, cushions of any kind being very difficult to dry when they get damp. It should be mentioned that when using modern single walled hives a crown board is used in place of quilts.

There is no true hibernation in the life of the honey bee; the bees merely get together in close formation between the combs, resting in a semi-dormant condition during the winter. If it can keep warm and dry this cluster remains unbroken as long as the cold weather lasts, but with every warm bright spell in the spring they break cluster and come out for a cleansing flight.

Spring Cleaning

In beekeeping the term 'spring cleaning' means the transfer of a colony to a hive which has been cleaned, disinfected and given a fresh coat of paint on the outside. This operation should be undertaken about the end of April on a warm sunny day between ten o'clock and noon.

I commence by giving the hive to be transferred a few good puffs of smoke through the entrance, then moving it to one side and placing the fresh hive on the old stand. The floor board is placed in position and carefully checked with a spirit level, leaving a slight slope forward so that moisture or condensation may drain out. The body box or brood chamber is then placed on the floor board, and as this is the part of the hive which is the permanent home of the bees it is important to make sure that it fits correctly. The top of the body box on which the lugs of the brood frames rest should be smeared with vaseline so as to facilitate handling and prevent the bees fastening them down with propolis, as they will do if this precaution is not taken.

Having proceeded thus far it is now necessary to give attention to the hive with the bees. Another puff of smoke is given through the entrance and the roof, lift, coverings and feeder removed. A slight puff of smoke through the feed hole should be given before the top cover is carefully taken off. Next push a flat scraper gently along the top of the frames removing any wax or propolis. Cover the bees on the brood frames with a square of clean cotton and beginning at the back of the hive roll the cotton forward as the work of transferring proceeds, care being taken not to kill any of the bees. Begin by loosening and gently lifting out the back frame, placing it in an exactly similar position in the fresh hive next the dummy board at the back. Repeat the operation with the other frames in turn till all are in place, and cover them with another square of cotton.

Before placing the queen excluder in position over the brood frames it should be smeared all over with vaseline. An excluder with long slots is best, and it will work more efficiently if correctly placed over the frames, that is with the slots parallel to them. As this is now the time to put a honey crate on the hive one should be placed in position on top of the queen excluder, having first smeared the under edges thoroughly with vaseline. The top of the crate is in turn covered with a square of oilcloth, smooth side down, the lift and warm packing placed back in position and the roof superimposed. Soon the remaining bees will have left the old hive which can then be well scrubbed out with hot water to which washing soda has been added, repainted and made ready for use when spring cleaning the next hive. During the entire process one has a good opportunity of examining the combs; noting the state of the brood; looking at the queen; removing faulty combs to the solar wax extractor and replacing with new ones.

Assuming that the colony transferred has been fed during March and April with honey syrup it will consist of a vigorous healthy stock of young bees, all ready for work when the call comes from the flowering orchards and assuring the owner of a rich harvest of fruit blossom honey. If, however, the colony has not been built up it will most certainly put to its own use the honey gathered from the fruit blossom, so that the precious harvest is lost to the parsimonious beekeeper; there will also be overcrowding at the wrong time with consequent risk of swarming. An already invigorated colony will, on the other hand, fill one or two crates of sections with honey, thus rewarding the beekeeper for his care and foresight. Every time a hive is opened after spring cleaning a watch should be kept for queen cells, but the beekeeper should proceed with deliberate haste so as not to slow up the work of the hive.

ROBBING

This is a trouble due to inexperience or carelessness and can turn an apiary of contented workers into an inferno, more especially at a time when there is a dearth of nectar in the fields. The utmost care should be taken to avoid the exposure of honey or wax about the hives, and every scrap of waste wax cappings or brace comb should be carefully placed in the solar wax extractor. If a partly filled comb has to be removed from a hive it should be taken indoors and put under a cloth, and if it is necessary to shake or brush bees from a comb this should be done over the opened hive so that any nectar from the comb can fall into it.

The entrance to each hive should be adjusted to the strength of the colony, and after the season is over only strong colonies should be retained as wasps can be a greater menace to a weak hive than robber bees, completely clearing it out in a short time.

Country people have a superstitious dread of the death's head hawk moth, *Acherontia atropos*, which is supposed to gain an entrance to bee hives in order to steal honey by imitating the high pitched soprano note of the queen bee. This moth, the largest species in the British Isles with a wing span of sometimes six inches, derives its name from the rather sinister escutcheon on the back of its thorax which has a striking resemblance to a skull and cross bones. As its tongue is comparatively short the nectaries of many flowers would be out of reach of this insect, and possibly this is why it visits bee hives in the hope of helping itself to honey.

As it is rare the moth is not often seen, and so it cannot be regarded as a serious menace to the beekeeper. Its association with the honey bee has, however, persisted from earliest times.

STINGS

Many persons no doubt would keep bees were it not for the natural fear of stings, but when the habits of the bee are thoroughly understood this fear disappears. The average beekeeper pays very little attention to a few stings on his fingers, and when bees are properly handled the number of stings can be reduced to a very low incidence. The moment a sting is received it should instantly be removed, otherwise it will gradually work itself into the flesh by muscular contraction discharging the contents of the poison sac. All beekeepers should wear a veil and have a good smoker always at hand. After one has been stung a certain number of times immunity is generally established and there will be little or no swelling.

The sting mechanism is very complicated and is situated at the end of the abdomen. It has a sharp two-edged dagger as tough as steel though pliable, toothed like a saw with the barbs placed as in a fish hook so that the sting becomes fixed

BROTHER ADAM EXAMINING BROOD COMB IN APIARY AT BUCKFAST ABBEY (*page 68*)

A. Worth

A COLONY ON ITS FOURTH CRATE (*page 105*)   *A. A. Lisney*

in the wound and breaks if an attempt is made to extract it. The sting is hollow and through it flows the poison.

In order to sting the bee must take time to obtain a firm hold on its victim before its weapon can be forced through the skin, after which it finds itself unable to move away and, if the sting remains in the wound, the bee cannot survive its loss. Usually the bee proceeds to walk round in a circle obviously trying to extract its sting, but only in very rare cases, and then only if it is left alone, does it manage to free itself. For many years, in order to save a life if possible, I have not attempted to remove a bee when stung on the arm, but only on three occasions have I known it to work itself free.

Remedies for bee stings are of little avail. It must be borne in mind that the poison is introduced into the flesh through a puncture in the skin so minute that the finest needle could not follow it up, and the flesh closes over so completely as to make it practically impossible for any remedy to penetrate the opening. Hot soapy water applied with a flannel does, however, give great relief to the pain.

This raises an interesting point. It has been said that being stung by bees will cure rheumatism, and people have kept bees in order to treat sufferers from this complaint; the insects are brought in small bottles to the patient and encouraged to sting the affected part. I wondered how bees would react in these circumstances and decided to try it out. With some difficulty I managed to take half a dozen bees from a hive in my handkerchief and brought them to the summer house. Lowering my stocking I spread out the handkerchief over my knee with the bees underneath and drew the stocking over it. The bees kept moving about, but made no attempt to sting and I waited some time before pressing on them, at which stimulus they all gave a slight prick but not one of them left the sting in my flesh; all I could see was a very tiny mark of no consequence. Over a period of some weeks I tried the same experiment, but the result was always the same and I came to the conclusion that the bees were only anxious to get free; they were prisoners and knew it.

As pointed out above, the bee, in order to sting, must first take a firm hold of its victim before it can force its weapon through the skin, and when bees are in fighting trim they draw blood. The only hope for rheumatic patients is for them to keep bees, that is if they believe that stings really bring about a cure.

After a bee has used its sting and torn itself away, a bundle of muscles partly enveloping the poison sac will be noticed above the surface of the skin, and the curious fact is that for some considerable time after the sting has been detached from the body of the bee these muscles will expand and contract in a rhythmic motion, thus forcing the venom through the sting into the wound. Even if a sting should come in contact with the skin after it has been extracted it may pierce the skin again, a function it can perform for fully twenty minutes after being disconnected from the bee.

The beekeeper must avoid crushing a bee while going through the hive, and the hands must never be jerked away; smooth movement from one operation to another is essential.

I have known of persons accustomed to handling bees for many years to become quite ill after one or two stings, and I do not in the least understand why this should be; probably the secret is to ensure that the hands and arms are scrupulously clean when handling bees so as to minimise the risk of infection. Mr. A. Worth, of Dorset, suggests that people whose blood stream is deficient in calcium appear to suffer more than others, and that a course of calcium treatment in late winter or early spring is beneficial.

As I mentioned before, the sting of the bumble or humble bee is most painful, but it is not barbed. Wasps also, which are very easily aroused to aggression, have stings without barbs and can use them over and over again. It is not necessary for a wasp to take hold of its victim in order to sting, and I have even known this insect to sting while on the wing.

THE SENSE OF SMELL

THAT the sense of smell is very highly developed in the bee is beyond question, and we owe to them the perfume of the flowers which, when they are ready for fertilization, send out fragrant waves so that all the bees can respond at once; they follow these waves to the source of supply and bring about fertilization. The flower gets much more from the bee than it gives, though this is a mutual service the importance of which cannot be over estimated; on this cöoperation between bee and flower hangs the life and health of mankind and the fruit and seed of the flower.

As I have previously pointed out, it is my belief that the scouts which lead a swarm each releases a scent from a tiny pearly crescent appearing on the last segment of the abdomen. In the same way the virgin queen, leaving the hive on her nuptial flight on a peerless morning, takes careful note of the hive and its surroundings before she turns and sweeps swiftly up into the blue vault of heaven. At the same time she releases this scent, which flows in delicate waves to reach all the drones in the neighbourhood drowsily lazing about in the orchards and gardens.

We humans are limited in our sensibilities and there is a vast field of sensations unknown to us, but it is an accepted fact that ordinary smell, the smell that affects our olfactory nerves, consists of molecules emanating from a scented body. The odorous matter dissolves and is diffused giving to the air the scent our noses register, in the same way as sugar dissolves in water and diffuses its sweetness. Smell and taste touch at some points; in both cases there is a contact between the material particles which give the impression and the sensory nerve endings receiving it. Without losing any of its substance a luminous point shakes the ether with its vibrations, and fills a circle of indefinite width with light. Smell in fact has two domains, that of the particles dissolved in the air and that of the sympathetic vibrations; the first we know, the latter belongs to the insect. The scent which attracts the bees to the flower is capable of spreading for miles around, so as to indicate to them that the nectar is being secreted in the chalices ready for fertilization.

Each hive has its own peculiar aroma, and no strange bee may enter a hive which is not its home. Drones are exempt from this rule, they come and go through all hives without hindrance as nature encourages cross fertilization.

If the beekeeper wishes for some reason to unite one colony to another, the distinctive scent of each must be confused by an entirely different one. To do this both colonies should be sprayed with a harmless perfume before being united, and smoked well during the operation.

When a hive has sent forth a swarm it is possible with care to restore it to the colony during the same evening, but confusion and disaster results if this is attempted on the following day, as in the meantime the swarm has developed a new scent of its own and will not be accepted by the bees remaining in the hive. We cannot comprehend these subtleties, or even dimly guess at the exquisite sensitivities of the olfactory and other receptive nerves of the bee.

It is well known that bees are highly sensitive to impressions they receive from people, and there are those who cannot approach to within reasonable distance of an apiary without provoking an attack by the bees and driven away.

I remember a curious instance of this. We had five children staying with us for some months and they were all able to play in the orchard near the hives, except one little boy. No matter where he went the bees singled him out for attack; he was not any more afraid of bees than were the other children and the reason remained a puzzle.

I was helping a friend remove a swarm from a hole in a rotten post, and we decided that after going through the hive from which it had emerged and cutting out queen cells we would put it back. As it was late in the season there was a full honey crate on the hive which I decided to remove and my friend took it away, as I thought, to the house, but as this was rather a long walk he had in fact left it on the ground some distance away from the hive. We got the swarm in the skep tying muslin over the mouth, and placed in a cool corner leaving all ready to hive it back during the same evening. It struck me that all the other hives had become unusually busy while we were working with the swarm and that the loud humming was abnormal, but I was absorbed with the work in hand and paid no particular attention. Presently, when all was tidy and the various odds and ends had been gathered together to take back to the house, my friend said 'I'll just collect the crate, I left it nearby.' My heart sank as I thought he had taken it to the house, and I then understood the ominous roar from the other hives. When we reached the crate it was unrecognisable, the bees from all the hives had been plundering, which is a different thing from harvesting as they become utterly demoralised. They had torn the wax cappings off in a frenzy to reach the precious liquid in the cells; wrestling and fighting they were using their stings to some purpose judging from the dead and dying lying around. The crate was almost empty and completely spoilt.

I have already referred to the theory of the dancing bees which signal where food is available so that the others can fly out and collect it; and that they report, in addition to the fact that there is food, the kind of flower in which it is found so that when the foragers leave the hive they are following a definite scent picked up from the antennae of the dancing bees. I find it difficult to accept this assumption, which to my mind over-elaborates a simple process.

Maeterlinck sought to prove whether or not a bee which he

had taken prisoner just as it was leaving the hive, had warned its companions of the existence and precise position of the source of nectar which it had lately been collecting. The result was negative; 'it is obvious,' he says, 'that if any verbal or magnetic communication had been made comprising a description of the locality and method of orientation, etc., I ought to have found in my study a certain number of bees so informed. I must say I only discovered one such bee. Was she following hints received in the hive, or was it a matter of pure chance? For my part I do not hesitate to say that it was pure chance.' He goes on to say 'In carefully repeating these experiments I always obtained the same result, *no flying directions were given in the heart of the hive by the busy harvesters.*' —The italics are mine.

Lord Lubbock saw one bee return fifty-five times in the course of a day to honey exposed on a table near the hive which stood indoors, not once did it bring a companion.

Bees are very sensitive to atmospheric conditions, and their reactions to pending meteorological changes can usually be relied on. If there is no activity around the hives when we have hopes of being favoured with a fine day beware, a disturbance of some kind is on the way!

DANCING BEES

I referred briefly in preceding sections to the phenomenon known as 'dancing' bees. As a lot has been written about this matter and many theories put forward in explanation, I feel that I must deal with the subject in more detail.

I have seen it stated that the life of the worker honey bee can be divided into three main periods, the first commencing when the bee emerges from its cell as a perfect insect and lasting until about the tenth day of adult life. During this period, it is said, the young bee looks after the preparation of the brood cells in the hive preparatory to the queen laying eggs in them, and undertakes other duties connected with the incubation and feeding of the larvae.

The second period begins on or about the tenth day of adult life and continues until about the twentieth day when, it is alleged, the worker has various domestic duties to attend to such as receiving, ripening and storing nectar brought into the hive by the foraging bees, and also cleaning and comb building. Towards the end of this period the young bee begins to fly short distances on its play flight during which time it learns the position of its hive, relative to neighbouring objects in the apiary. About the twenty-first day the worker bee commences the third and final period collecting water, pollen and nectar. This to my mind over elaborates the economy of the hive bee and, in my experience from studying a hive of black bees which I had requeened with a pure Italian queen at the beginning of a honey flow, the young pure bred bees began their play flight when just over a week old and when ten days old were out harvesting.

One bee does not on its own initiative find the source of nectar, but all the bees in the neighbourhood receive the scent of the flowers at the same time. How, therefore, can a single bee know before all the others that a rich harvest is ready for gathering, and by a tail-wag dance and figures of eight pass on the information to a hive of from 40,000 to 50,000 others? We are not told how the first bee is supposed to get its information as nothing in the smooth running of a hive is left to chance.

I have for many years noted these so called dancing bees in the hive. Sometimes a single bee is seen to be quivering with excitement, spinning on the frame and very amusing to watch. These antics are often kept up for some time regardless of the fact that the frame is being held in the hands and, as far as I have observed, none of the other bees take the slightest notice of their eccentric companion wholly absorbed in its own wild spinning. So rapid are these gyrations that it is impossible for one to distinguish the outline of the whirling bee, looking very much like a humming bird hawk moth hovering over a flower sipping nectar; it is a mad ecstasy of *joie de vivre* without a logical explanation.

Much more often seen is an entire frame of bees absorbed in a rhythmic motion, but there is the difference in that while the single bee spins rapidly, the frame of bees seems to be agitated as a whole by muscular movements and twitchings of their wings which do not actually vibrate, nor do the bees alter their position on the frame. It is always over the brood that this collective phenomenon occurs, and only when the hive is open in weather which is cool for the time of year.

It is, of course, well known that bees can regulate the temperature of the hive by muscular contractions, movements of the body and twitchings of the wings. These rhythmic motions I have come to associate with cooler conditions in late spring and early summer, when it is so necessary that the temperature of the hive should be maintained at 95° Fahrenheit over the brood and not drop to a level which would chill them.

To sum up, I feel bound to repeat that I cannot accept what to my mind are theories to explain phenomena which after all are the simple and natural functions of the hive. It seems to me utterly fruitless to endeavour to explain such episodes in the insect world in terms of human experience.

NUPTIAL FLIGHT

When the young virgin queen has emerged from her cell and assured herself that there are no rivals of which to dispose, she walks around the combs for several days. The workers pay no heed to her; she must take her maiden flight and return as the fertilised mother of the hive before they pay close attention to their queen.

It is of the utmost importance that the nuptial flight should be successful, and it must be brief because of perils in the air; that is why the virgin queen awaits a calm day and a flawless sky. She seems to be conscious of the glory of the day and hour and those magic wands, her antennae, play their part in probing out the meteorological conditions prevailing on that auspicious occasion. The instinct of the queen in preparing for the nuptial flight with scrupulous care leaves us lost in wonder.

One day she crosses the threshold of the hive and makes her way through the busy crowd of workers emerging and returning. Hestitating for a moment she strokes her head and the flexible joints of her antannae with her front legs, and passes her hind legs several times across her abdomen by way of giving it a careful brushing.

It is noon and the drones have long since left the hive, mingling with others from neighbouring colonies they buzz and whirl around the gardens and orchards. Suddenly from the hive entrance the queen emerges, turns about, and with her wings hooked together takes to the air. After hovering for some time with her head directed towards the hive, she moves in ever widening circles taking careful note of the hive and its surroundings. These precautions are necessary for obvious reasons; the risks are exceptional, and her life and the lives of many workers would be lost by the slightest mistake. The countryside is drowsing in the noontide sun and there is not a breath of wind; the azure dome of heaven stretches into illimitable space; everything is propitious. Once the young queen has determined her position she suddenly turns away from the hive, and facing up into the blue sky flies swift as an arrow disappearing from sight. She climbs higher and higher to where there is no sound and where no other wing beats stir the air. She leaves in her wake a subtle fragrance and, as I mentioned previously, I consider it reasonable to assume that this scent is released from a gland on the last segment of her abdomen.

We see the reason now for the generosity with which nature has endowed the drones, for they have a sense of smell of the very keenest and eyes peculiarly adapted for sweeping the heavens. They possess a massive thorax and muscles of great strength so that their wings can cleave the air in whirlwind circles. *It is only after the end of a strenuous flight that the drone is equal to his task of mating.*

On many occasions I have been up the bee walk on a perfectly glorious morning when a young queen has ventured on her nuptial flight. The drones from all over the garden and orchard soon detect her subtle fragrance and discover the fugitive, so that the air is filled with a deep musical diapason as thousand upon thousand of drones launch themselves into the air in full flight after her. As they sweep upwards their low pitched harmonies are thrilling to hear.

It is only the strong and well nourished that can keep up the swift flight the weaklings and decadent being left behind, outstripped by the more enduring. The frenzied drones press on in the mad race, gradually gaining on the speck of gold lost in the blue dome of air. More and more are left behind; many overcome by exhaustion perish on the way. Thinned out to no more than a few they press on, air being drawn into their tracheal tubes so that the air sacs, empty and collapsed during repose, are full and dilating. Soon the goal is within reach and the most vigorous of all makes a gigantic effort to outstrip his last few rivals, sweeping forward to join his queen in matrimony. The winner of this mad race is not necessarily a drone from the same hive as the queen, as one of the laws most favoured by nature is cross fertilization.

Some have claimed to have witnessed the termination of this dizzy pursuit in the consummation of the nuptial flight, but—

> 'Who hath eyes so strong or so keen
> That they can follow the flight of the queen?'

The consummation takes place far beyond the range of human sight, but I have been thrilled to have witnessed as far as is possible the wedding flight on so many occasions.

The epilogue to this, one of the many marvels of nature, is that hardly has the drone reached his bride for one brief moment than, from the height of heaven, the now unfortunate victor over all his rivals falls back to earth a forgotten corpse. So ends the sublime adventure.

ARTIFICIAL INSEMINATION AND OTHER ABUSES

The idea makes one shudder, and the introduction of the subject of artificial insemination into beekeeping is entirely unwarranted and unnecessary, as no improvement on nature in this sphere can be effected.

I have read recently of an attempt over many years to build a honey comb as good as that produced by the bee. All efforts were, however, crude and the bees still had to use as much wax to line it as if they had made the foundation themselves. The idea behind these experiments was, of course, that more honey would be produced as the bees would not have to use so much in making wax.

If, therefore, man has failed to build a better honey comb than the bee, it is unlikely that any attempt of his in selective breeding by means of artificial insemination can improve on nature.

For centuries man was the greatest enemy of the bee in spite of the fact that he feared its formidable sting, and in order to possess himself of the honey with the least discomfort to himself he destroyed the most fruitful skeps and spared the least vigorous.

As each generation of honey bee came and went their vitality was being gradually drained away, depriving them of a healthy resistance to disease. Were it not for the fact that they were able to mate as God meant them to, I wonder what awful calamity might have overtaken our hive bee. On the nuptial flights of those tragic days of weak stock the undersized and undernourished drones were weeded out, with the result that a certain balance was maintained in spite of man's want of thought. It fills one with horror to think of man presuming to imagine that he can improve the honey bee by any unnatural interference.

It is hardly to be wondered at, therefore, that in the course of time the terrible scourge of Isle of Wight disease swept through the country from south to north, leaving a trail of dead and dying in its path. Thanks to Miss Elsie J. Harvey and Dr. Rennie we know that this destructive disease is due to the invasion and lodgement of a fertile female mite *Acarapis Woodi* within the trachea of the hive bee. The mite is blind and must reach its host by scent, but if the scent is confused by another more potent the mite has no way of discovering its host and dies of starvation.

As a precautionary measure against invasion by the mite it is a good plan to soak some cotton wool in methyl-salicylate and put it on the floor board or under the top quilt in November, renewing the dose about every six weeks subsequently. No measured dose of methyl-salicylate is necessary as it is not lethal to bees, neither does it stupify them thus leading to robbing by other colonies.

Exploitation of our bees for the last drop of honey still takes place. In a paper some time ago I saw the picture of a man bending over one of his hives, with the caption underneath to the effect that after a very profitable harvest this generous beekeeper was feeding his bees with sugar syrup. He had obviously removed all the honey they had laboriously gathered, and was giving them a very poor substitute in return.

Must the bee still be sacrificed to our greed? It takes ten thousand journeys from the hive to the field and back again to fill a one pound section of honey, and ten weeks off and on is the normal duration of the honey harvest. How often do the conditions obtain that make the gathering of the harvest possible? The flowers must be fully open yet not mature ; there must be soft clouds to veil the sunlight; moist air from the south and a calm day; yet during a normal honey flow a populous hive will produce 156 pounds of extracted honey.

It takes from twenty-five to thirty pounds of honey successfully to winter a strong colony of bees, and as such a colony will easily give a hundred pounds of honey it is only fair to return to them sufficient for the winter months. As a rule, when hives are worked for section honey alone the bees store an appreciable amount in the brood nest, but if worked only for extracted honey all their stores are to be found in the supers. If bees are given sufficient stores for the winter candy is not necessary, and honey goes much further than sugar syrup. The bees are able to generate a much higher temperature in winter with honey than with syrup, and a weak hive has a chance of coming through the cold months if supplied with honey when it would perish on syrup.

Nothing, therefore, can possibly take the place of honey as food for bees, especially at a time when brood is being reared. Deprived of honey for their winter stores it is very difficult for them to prevent the general temperature of the hive falling below 57° Fahrenheit, at which temperature bees cluster. Foul brood, which is probably a deficiency disease, and all those ills which lie in wait for the undernourished can be expected to pass by the strong colony; no longer need scientists have difficulties trying to discover cures when prevention is so simple. In proof that I practice what I preach, I might add that I have never experienced either Isle of Wight disease or foul brood.

I honestly believe that God created the bee and the flower so that in harmony with each other, fruit, seed and honey is provided for man, otherwise it is difficult to understand why the bee stores far more honey than the requirements of the hive justify.

What an empty world it would seem without the bee family; no perfume, no harvest of fruit or seed, and everything sterile. Every time we eat an apple, pear, or plum let us remember that these are the fruits of the hive as much as the honey. The bee carries on from one generation to another fulfilling the work ordained for it from the beginning, so let us see that it gets its full share of the honey harvest and not be guilty of anything that would upset this exquisite balance of nature.

As for man, from time to time leaders of various schools of thought have put forward ideas for the improvement of the human race. Contrary to the bee, man has his own free will and is more or less responsible for the conditions under which he lives; the choice of right or wrong are his and he was born for a supernatural destiny. He only knows what perfect freedom is when bound to his Creator.

THE MUSIC OF THE BEES

Charles Butler, writing in his *Feminine Monarche* published in 1609, gives some staves reproducing the song of the bee; the Melissomelos or Bee Madrigall. According to him, the bees at swarming time, with many harmonious notes and pauses in between, go solemnly round the hives to give warning to all the company. This music, he says, must please and delight all who listen to it, and be most sweet and pleasant to the young princess herself; the queen pipes a clear soprano, while the young princesses take a higher key. He loved to accompany his bees on a stringed instrument. Undoubtedly the hive has a voice composed of thousands of imperceptible murmurs, forming a deep toned muffled song.

If the queen is removed from a hive there will be a cry of lamentation and hopeless sorrow in a very short time. This sound is shrill in tone and increases in volume as the grievous news is transmitted through the colony. When she is returned to the hive, immediately a note of joy can be detected.

The note of the bee varies with the seasons except when they act as water carriers; then the note is always plaintive as they wing from the hive on this so necessary errand full of many dangers.

After the months of ice and snow and stinging north winds have gone and the earth grows warmer, the bees feel the stir of the bursting buds as the leaves emerge from their sheaths.

With the renewal of life on the earth there is also an awakening of vitality within the hive. Already many pupae await emergence, and as a gleam of warm sunshine falls on the threshold of the hive thousands of bees of all ages pour from it. Their melodious hum in springtime as they take this airy dance transcends all the musical notes of the unfolding year, each bee contributing to the harmonious whole; it is a song of joy and hope and the music is infectious, giving rise to a response in the hearts of those who hear it.

When the honey flow comes, which it does swiftly, the note rises clear and pure as the tumultuous horde shoots from the hive at the call of the flowers, only to drop again to a pulsating minor key as the bees return laden with nectar. For many days this drowsy rhythm is heard and is very soothing to the ear.

It is nine o'clock in the morning, the sun has swept the dew from the grass and there is a fragrant heat as we take a turn up the bee walk. Suddenly a familiar note strikes the ear, the music of a swarm. They come out almost walking till they reach the end of the alighting board, then fling themselves into the air with wild abandon so that presently a choir of 40,000 bees or more hover over the orchard or garden. It is a glorious melody, with the basoon notes of the drones rising and falling in the triumphant happy chorus; the sweetest music it is possible to imagine.

Thus the summer passes and the shorter days draw near. The precious stores must be guarded and so all bees now become warriors; their note is sharp, angry and menacing as they hurl themselves at anyone coming too near the hive. Even the most docile races will draw blood at this time.

As autumn approaches and chill gusts of wind give a foreboding of the long dark winter days, this irritable note changes to one of pathos and deep sorrow. Summer is gone, the bees have built up stores to provide for winter when they must pass the time looking after the queen and keeping the temperature of the hive constant. As we have been with them all through the year we understand when we hear the 'sad autumn winds that wail as they pass o'er trees bare and leafless and the dry withering grass.'

THE TRUMPETING OR PIPING OF THE QUEEN

In a hive where the old queen has left with a swarm and been hived elsewhere, queen cells are to be found in various stages of development. If one emerges and the remaining bees decide she must head what is known as a first cast or secondary swarm, they will not allow her to interfere with the other queen cells, and on a still warm evening the trumpeting of the rivals from within these cells can be distinctly heard. This is a shrill note, clear and defiant in a soprano tone ; a challenge to any rival for battle. It sounds like 'z-e-e-p, zeep, zeep,' that is a prolonged note followed by several much briefer, each rather shorter than the preceding. It is of great interest to hear these strange notes ringing out in challenge.

On an occasion some years ago when several cages containing queens had arrived from Italy and were left together on a table, a chorus of trumpeting or piping could be heard some distance away. It is not definitely known how these notes are produced.

THE MASSACRE OF THE SUITORS

When the sun begins to climb its way to the zenith, the bees feel an urge to feed the queen in order that she may produce the eggs and larvae in anticipation of the harvesters needed as the months unfold.

So it is also in July when the sun has begun to descend in the heavens, and there is a definite slackening of the honey flow, that the breeding of drone larvae is discontinued altogether; the glorious days of abundance are past. The food given to drone larvae differs from that fed to the workers, containing more albumen and less sugar. Of stronger build a drone weighs twice as much as a worker bee, and to rear a

A LARGE SWARM (*page 105*)   *A. A. Lisney*

THE BEE WALK IN WINTER (*page 110*)   *C. Checkley*

thousand drones would take as much nourishment as would rear fifteen hundred workers.

Drones never feed on matured honey but require fresh nectar, which explains why the birth rate of drones rises and falls with the abundance of the harvest, and also why they are not tolerated in the hive after the honey flow has ceased.

The workers remain more or less indifferent to the drones up to the end of July or beginning of August according to the season, and they are allowed to fly and feed as a heedless carefree crowd. Then there comes a sudden change; a feeling of constraint. After their play flights in the sunshine the drones are stopped and interrogated before entering the hive. The workers are rough with them and they, foolish and bothered, spoilt princes of the hive, cannot understand. The resentment increases and spreads through the colony so that the drones are hustled, followed up and seized from every side, as many as five workers to every unfortunate drone. Each pulls himself away by sheer weight only to be seized again by as many more workers who, with their powerful mandibles, saw off his wings and drag him to the edge of the alighting board where they all topple over together. Poor fellow, he cannot fly and now half starved and worn out with fatigue, broken and battered, he becomes a prey to wasps who carry off as much of his body as they can to feed their own queen larvae.

When night falls the alighting board is strewn with helpless outcasts. They no longer attempt to cross the threshold of the hive and no wonder, for the sentinels are on guard in full force; when morning dawns these outcasts will all have perished with cold and hunger. Alas, the bees must protect the honey which is the daily bread of the hive, and so the massacre becomes a necessity to preserve the race.

It has been asserted that worker bees use their stings on the drones, but in all my experience I have never seen them do so. There is no doubt, however, that the workers deprive them of

food as autumn approaches and that the death of the males is
an indispensible condition of survival to the colony.

HONEY AS A FOOD AND FOR HEALTH

During the centuries previous to the advent of cane and beet
sugar, honey was the only concentrated sweetening agent in
the world. The introduction of cheap sugar has made a great
change, and the consumption of sweets has increased beyond
belief; in the United States alone the consumption of sugar per
head per year is eighty-five pounds. We should pause to
consider the probable effect of eating so much crude sugar,
and it is a serious question as to whether the present generation would not benefit if honey were restored to a more
prominent place in the diet.

Honey is more easily assimilated than any of the other
carbohydrates ; crude sugar having a tendency to produce
irritation in the digestive tract and sometimes causes indi-
gestion. Honey on the other hand is free from this fault as it
does not require digestion at all, its sugars being already
inverted and, therefore, ready for absorption into the blood
stream without change.

From time immemorial the virtue of honey as a remedy has
been fully recognised, and even in these days of patent
medicines many people think of honey only as a remedy for a
sore throat. The chief virtue lies in its high nutritive value and
it will fortify the body in conditions of grave illness as it does
not, like sugar, ferment in the digestive tract. Where there is
ulceration for instance, it will nourish and stimulate without
any unpleasant reaction. Minute quantities of valuable
minerals, which cannot be too highly appreciated, consist of
iron, phosphorous, lime, sodium, potassium, sulphur and
manganese; there are also traces of oils, gums, essences
natural to the flowers, and certain vitamins.

Many of the minor and major ailments of life are now being
traced to deficiencies of vitamins and minerals, and honey is
rich in both these essentials. It is a mild laxative, a bone and

body builder, and invaluable to those with a weak digestion. There is no better food for children and their craving for sweets should be supplied from nature's own laboratory, the nectaries of the flowers.

Dr. C. J. English has written 'In diseases of the respiratory tract, honey was given for its value upon the heart muscles. When patients neared or were in the crisis of pneumonia, a very marked effect was noted if honey was given rather freely. I found it also noteworthy that these patients would tolerate large quantities—several ounces of extracted honey per day. The benefit was so apparent that it became routine practice in conjunction with other treatment. No other food or heart stimulant had such a lasting effect upon the myocardium . . .'

All natural honey contains pollen from the blossoms made up of grains which, though exceedingly minute, are all different and characteristic of the various flowers from which they come. It is comparatively easy, therefore, to ascertain the source of a sample of honey by examining the pollen in it.

Foreign honey can often be detected when sold as British owing to the presence of pollen from flowers not found in this country. Native honey is far superior to that generally imported, and is well worth the extra cost.

It is of interest to know the colour of honey from various sources:—

SYCAMORE—very dark. Showing as darker streaks in the early sections and supers.

FRUIT BLOSSOM—dark with sometimes a greenish shade and strong flavour.

HAWTHORN—is a very thick amber coloured honey of almond flavour.

SAINFOIN—lemon yellow with a mild flavour. Comb bright yellow.

CLOVER—pale amber with a mild delicious flavour. Comb white.

LIME—rich amber and of a thick consistency with a delicious flavour.

HEATHER—dark rich brown or purple. Comb snow white.

THE SWARM THAT WAS TURNED BACK

According to Maeterlinck, 'Once on the wing a swarm cannot be turned back,' which prompts me to give a rather detailed account of a sequence of events disproving this theory.

A colony of bees which had been received from the South of France made fine progress, and towards the end of June they had almost filled the third crate and did not appear keen on swarming; I was of course keeping an eye on them. One morning I was away from home, and on returning in the early afternoon I took my favourite turn up the bee walk. As I passed this particular hive there, hanging from the branch of an apple tree nearby, was a huge swarm; it looked lifeless so still were the bees. I guessed from their motionless close formation that they must have come out early that morning.

I hastily procured a skep, approached the lovely cone of bees and holding the inverted skep underneath put out my hand to catch the branch intending, with a quick dexterous jerk, to precipitate the bees into it. At that very second, however, the swarm fell to pieces, rose into the air, hovered for a moment and disappeared over a wall.

I thought very swiftly and decided that here was a golden opportunity for my little son. I collected a watering can full of water and a hand syringe which were always ready for just such an emergency and, calling to my son, 'fly,' I said, 'a swarm is crossing the field'—at the same time giving him the can and syringe—'turn them back if you possibly can by directing the water up in the air ahead of them.' He was gone almost before I had finished speaking and collecting a few things which might be required, I followed. Indeed we all followed, and even the cows in the field joined in a mad race as we hurried along. I noticed in the distance that my little son had stopped and was using the syringe for all he was worth, and that the bees had paused like a dark cloud in the air. I shall never forget the excitement of that moment as we caught up with them. What would the bees do next? Perhaps wheel left or right, in which case we should surely lose them as there was a railway cutting a short distance away on one side of the field and a main road on the other. The bees hovered in the air apparently trying in some mysterious way to solve a difficult problem and we all stood around gazing up at them, the cows in turn gazing at us with absorbed attention. At this point the cows suddenly created a diversion. The bees, evidently believing that the cattle had something to do with the miscarriage of their plans, sent some warriors down on them. There was a terrific stampede as the cows scattered to the four winds tails aloft.

Then the magic of some subtle command travelled like lightning through the cloud of bees as, with a single motion they wheeled into close formation and headed back from whence they came, all the family hastening after them. We crossed the wall at record speed to where the swarm was circling around and falling like rain back on the apple tree, and we watched them breathless as they formed into a cluster on the exact spot from which they had departed a short time previously. I syringed the bees lightly and presently had them safely in the skep with muslin tied over the mouth; I left them to hive later. This swarm weighed six pounds, and as there are about 4,000 bees to the pound there must have been approximately 25,000 bees altogether.

I examined the hive which the swarm had left removing the queen cells and also transferring a few frames of brood to a weaker hive, replacing them with empty frames for the queen to lay in. I also put on a fourth crate as the three crates in the hive were full. It was late enough in the year for these few precautions to prevent further swarming.

APPENDIX

Beekeeper's Calendar

January

This is a very anxious month for the beekeeper. If there has been continuous cold with temperatures at freezing point or a little above, and not even a 'pet' day to give the cluster a chance of breaking up and moving to a fresh feeding place, the beekeeper has cause to feel anxious. Much heat is lost in the dispersing of the bee nest and a colony may die of starvation with plenty of sealed stores a few inches away, but the careful beekeeper has foreseen this possibility by placing a flat cake of candy over the centre of the brood nest. A cake one and a half inches thick and six to eight inches square is a safeguard against starvation.

February

It is a good thing to look under the cover next the brood nest during the month on a fine day in order to determine the food situation. If everything is satisfactory leave alone, but if the candy is finished place a smaller cake in position with as little disturbance as possible. The queen begins to lay this month in the cosy centre of the bee nest.

On springlike days if the air is warm enough, the bees will come out on a cleansing flight. There are always the water carriers, who often have to battle against the elements in order to carry home the precious drops which, mixed with pollen, is fed to the larvae.

March

The temperature of the hive is rising. One can judge how busy the bees are by all the waste matter carried out and thrown from the alighting board; cappings of honey and brood combs and alas, the dead bodies of bees that have

107

perished with cold on the outside of the cluster. Sometimes a large piece of grease-proof paper from the candy emerges with two bees actively engaged to deal with it, a bee struggling with each end; bravely they take the air carrying their cumbersome burden to a safe distance. What else is coming out now, a large pin!! I remember I had used it to fasten a queen cell on the comb. It is decidedly awkward, but the bee holds it by the head and manages to remove it.

Pollen and nectar are being carried in with great excitement; the hazel catkins are out and the gorse is on fire with golden flowers.

Early in the month the beekeeper should place the feeders in position and every evening give a little honey syrup.

APRIL

Open the slides of the entrance a little more. What a wealth of nectar and pollen there is for the taking from the blossoms of flowers and shrubs such as *Berberis Darwinni*, hyacinths, the humble arabis and *aubretia*, grape hyacinths, *Daphne* and many others. The brood nest is rapidly increasing, and the evening feed of a little honey syrup builds up an enormous colony. Weather permitting spring cleaning begins at the end of this month, and all hives should be given a crate or super.

MAY

Boards that slope from the alighting board to the ground should be placed in position now and the slides of the entrance may be fully opened. Fruit blossom is coming in and nectar from all sources is abundant; there is an atmosphere of kindliness and goodwill. One can with confidence open the hive between ten o'clock and noon on a glorious day in order to form an accurate assessment as to how things are progressing in the hive. Learn to recognise brood in all its stages, from egg to perfect insect. If any queen cells appear cut them out, unless it is desired to rear young queens. When a crate or super is little more than half full place a fresh one underneath, keeping pace with the requirements of the hive and not forgetting the vaseline.

JUNE

Colonies will be at their best this month. Keep a little note-book for jotting down items of importance such as the most productive hive, the colony that shows little or no inclination to swarm, or anything that should be remembered. Give crates and supers as necessary so as to prevent overcrowding.

If a hive swarms and the intention is to increase the apiary, place a fresh hive on the site of the one from which the swarm has departed and fill it up with frames of brood foundation, placing the queen excluder in position under the crates or supers. Cover all up as before and hive back the swarm the same evening, in the meantime moving the old hive to a fresh stand some distance away. The bees will all fly back to the old stand where the fresh hive has been placed, and settle down to store honey. Cut out all queen cells in the old hive except the best unsealed one, and after about a week go through the hive again leaving only the same queen cell which will have been sealed.

All young queens should be mated this month.

JULY

Half way through this month the honey flow is usually over for the year. All crates and supers should be removed and all honey extracted, allowing the bees to retain for themselves any nectar they may be able to gather from this on. See that all hives have a queen not older than the second year. If necessary re-queen from a reliable source according to the instructions accompanying the travelling cage, or from one's own stock.

AUGUST

After the honey has been extracted sterilise all appliances with boiling water, carefully drying and then oiling them before putting in store. Close the slides of the entrance to four inches, and give a little stimulative feeding to keep up brood rearing as long as possible so as to ensure a vigorous population of young bees for the winter. The bees will be busy this month gathering propolis and rendering the hive as damp and wind proof as possible.

SEPTEMBER

After the stimulative feeding in August rapid feeding is now necessary, remembering that it takes thirty pounds of sealed stores to winter a colony of bees. Close entrances to two inches as wasps will have become active.

OCTOBER

All hives should have a large cake of candy placed in position over the centre of the brood nest with the hessian cover on top, and then all packed warmly down with a square of quilt, old blanket, or carpet. Hives should be made secure against storm and wind, and particular care should be taken to render them mouse proof by closing the entrance sufficiently.

NOVEMBER

The beekeeper can take a rest from his work among the hives as the bees will have settled down for the winter.

DECEMBER

Silence reigns supreme.

INDEX

www.ingramcontent.com/pod-product-compliance
Lightning Source LLC
Chambersburg PA
CBHW050130290326
R18043500002B/R180435PG41927CBX00019B/1